Supplement Issues of
Zeitschrift für Kristallographie

No. 1
The Non-characteristic Orbits of the Space Groups

R. Oldenbourg Verlag München 1984

The
Non-characteristic Orbits
of the Space Groups

by

Peter Engel, Takeo Matsumoto,
Gerhard Steinmann, Hans Wondratschek

Supplement Issue No. 1

Zeitschrift für
Kristallographie
International Journal for Structural, Physical,
and Chemical Aspects of Crystalline Materials

Editors-in-Chief

M. Buerger, Cambridge, MA · S. Haussühl, Köln
H. G. von Schnering, Stuttgart
with the assistance of W. Hönle, Stuttgart

R. Oldenbourg Verlag München 1984

PD Dr. Peter Engel, Universität, Laboratorium für Chemische und Mineralogische Kristallographie, Freiestraße 3, CH - 3012 Bern

Professor Dr. Takeo Matsumoto, Department of Earth Sciences, Faculty of Science, Kanazawa University, 1-1 Marunouchi, Kanazawa 920, Japan

Dipl. Min. Gerhard Steinmann, Institut für Kristallographie, Kaiserstraße 12, D-7500 Karlsruhe

Professor Dr. Hans Wondratschek, Institut für Kristallographie, Kaiserstraße 12, D-7500 Karlsruhe

Gesamtherstellung: R. Oldenbourg Graphische Betriebe GmbH, München

ISBN 3-486-28891-1

Contents

List of Tables

1. Preface

In crystallography, symmetrically equivalent objects play a fundamental role. Equivalent faces are the basis of morphological studies. Equivalent directions are morphologically important in the form of equivalent edges; moreover, they are the starting point of classical crystal physics of anisotropic media. Equivalent points are considered in crystal chemistry and in crystal structure analysis as the centres of equivalent atoms or molecules, as voids of equal surroundings, or as loci of equal symmetry.

This monograph is devoted to the study of the sets of all points which are equivalent under the symmetry operations of a space group. Such a set is called a crystallographic orbit. Explicit tables of crystallographic orbits are available and in permanent use, above all the International Tables for Crystallography, Vol. A (Hahn, 1983), referred to as IT 1983 here. For each space-group type, in IT 1983 the crystallographic orbits are classified according to Wyckoff positions, i.e. to the positions and site symmetries of their points.

The crystallographic orbits of different Wyckoff positions of the same space group may belong to different phenotypes. Moreover, even the crystallographic orbits of the same Wyckoff position may differ considerably in their appearance and in their apparent symmetry, for examples cf. section 2.3. Regarding this, a crystallographic orbit may be called characteristic if its intrinsic symmetry (eigensymmetry) is that of the original space group, and non-characteristic if it displays a higher symmetry.

More than fifty years ago Paul Niggli raised the problem of finding all non-characteristic crystallographic orbits for each space group, but up to now all attempts in achieving this aim were unsuccessful. The tables presented in this monograph solve the problem. The essential part is the Main Table of the non-characteristic crystallographic orbits. The accompanying text consists of chapters on basic definitions, comments to the Main Table, the methods used to derive and check the tables, applications of the tables and historical remarks.

Originally the data of these tables were planned to
be part of IT 1983. However, when the preparation of
IT 1983 began no systematic method to derive the
complete set of non-characteristic orbits was available. In the seventies a systematic procedure was
developed but the methods of calculations used at
that time were time consuming and progress was slow.
Nevertheless, the first results gave stimulus to a
more thorough examination of the theoretical basis
and an improvement of the methods. These improve-
ments enabled us to expedite the calculations by
computer and by hand considerably. The final
solution was achieved through the co-operation of
the authors, using a combination of theoretical
results, hand calculations, and data from pertinent
computer programs.

Although much care has been taken, the compilations
by hand and the insertion of part of the computer
results by hand make occasional errors and misprints
unavoidable. In spite of the extensive tests even
systematic omissions or mistakes due to errors in
the computer programs are not impossible. The
authors would be grateful for notification of any
such errors, omissions, or misprints detected.

Applications of the data may be foreseen primarily
in the fields of theoretical crystallography, crystal chemistry, and crystal structure analysis. The
authors will be satisfied if this volume stimulates
further investigations in these fields, and if the
contents help in getting a better overview over the
relations between different crystal structures and
different crystal-structure types.

2. Introduction

In this chapter some fundamental definitions are given in order to explain the terms used in this volume. The International Tables for Crystallography, Vol. A (Hahn, 1983), abbreviated IT 1983, are considered as the basis. The nomenclature and the conventions to be found there are taken as standard.

2.1. Crystallographic orbits

A crystallographic orbit $O_G(X_0)$ in n-dimensional Euclidean space is the set of all points X_i which are equivalent to the point X_0 under all motions of a space group G,

$$O_G(X_0) = \{X_i, \ X_i = g_i X_0 \mid g_i \in G\}.$$

Example. The crystallographic orbit of the point $X_0 = 1/8, 1/3, -1/8$ (referred to the conventional coordinate system) in a space group G = Pma2 is the set of all points with coordinates

    1/8+m, 1/3+n, -1/8+o;   -1/8+m, -1/3+n, -1/8+o;
    5/8+m, -1/3+n, -1/8+o;   3/8+m, 1/3+n, -1/8+o;

    m, n, o any integers.

The point X_0 is called the representing point of the crystallographic orbit. Any point of $O_G(X_0)$ can be taken as the representing point, in IT 1983 the standard representation is restricted by the conditions $0 \leq |x_0|, |y_0|, |z_0| < 1$ for the coordinates of X_0.

The space group G is called the generating space group. The orbit is called crystallographic because of its discrete nature and its periodicity. Discrete means that around each point $X_i \in O(X_0)$ there exists a spherical region of radius $r(X_0)$ which contains no other point of the set. Crystallographic orbits have been considered frequently under different names, e.g. regular point systems or point configurations.

The set S_0 of all symmetry operations $s \in G$ such that $sX_0 = X_0$ is called the site-symmetry group S_0 of the point X_0 with respect to G. The site-symmetry groups of all points $X_i \in O_G(X_0)$ are conjugate subgroups of G, i.e. $S_i = g_i S_0 g_i^{-1}$ with $X_i = g_i X_0$.

If the site-symmetry group S_0 is the identity operation 1, the crystallographic orbit is called a general orbit. Otherwise, if $S_0 > 1$ it is called a special orbit.

The crystallographic orbit in the above-mentioned example is general, the crystallographic orbit of $X_0 = 1/4, 1/8, 1/2$ of the same space group is special because $X_0 = mX_0$ holds, with $m \sim 1/2-x, y, z$. The site-symmetry group S_0 consists of the symmetry operations 1 and m, it is of order 2 in this case.

2.2. Wyckoff position, Wyckoff set and type of Wyckoff sets

Crystallographic orbits are usually classified according to their site-symmetry groups in the generating space group G. Let X_0 be a point and S_0 its site-symmetry group. The set of all points X_1, for which S_1 is conjugate to S_0 in G, is called the Wyckoff position $W_G(X_0)$ ("Positions" in IT 1983) of X_0 under G. The different "Positions" of a space group G are designated by different letters, the Wyckoff letters.

Example. The Wyckoff positions of the above-mentioned space group Pma2 are characterized in IT 1983 by the multiplicity, the Wyckoff letter and the oriented site-symmetry symbol: 2a ..2, 2b ..2, 2c m.., 4d 1. In this monograph only the type of site symmetry is stated and thus the Wyckoff positions are characterized by 2a 2, 2b 2, 2c m, and 4d 1 or simply by 2a, 2b, 2c, and 4d. In addition, a representing manifold is printed in most cases, in the example 0,0,z; 0,1/2,z; 1/4,y,z; and x,y,z; respectively, or, e.g. 4b $\overline{1}$ 0,0,1/2; in Pccn.

One of the major achievements in crystallography is the classification of the infinitely many symmetry groups (space groups) of all possible ideal crystal structures into a finite number of classes, the so-called space-group types (often called "space groups" too). This is done in the following way.

Let $\mathbb{A}$ be the group of all affine mappings and $\mathbb{A}^+$ the special subgroup of all affine mappings with positive determinant. Two space groups G_1 and G_2 are said to belong to the same space-group type if there is an affine mapping $\alpha \in A$ which maps G_1 onto G_2: $G_2 = \alpha G_1 \alpha^{-1}$. According to whether there exists an α in A or in A^+, 219 affine or 219+11=230 special affine (or crystallographic) space-group types are distinguished.

The concept of Wyckoff positions of space groups has proved very useful. It cannot, however, be transferred to space-group types, as the Wyckoff positions of a space group G may be permuted to a certain extent when G is mapped onto itself or onto another space group G' of the same type. For an example see IT 1983, p. 725. In order to find a concept for space-group types, corresponding to the Wyckoff positions for space groups, "Wyckoff sets" will be introduced now. This can be done with the aid of normalizers.

The set of all affine mappings α which map a space group G onto itself, $G = \alpha G \alpha^{-1}$, forms a group which is called the affine normalizer $N_A(G)$ of G. The affine normalizers of all space groups are well known. They are listed, e.g., by Burzlaff and Zimmermann (1980), Billiet, Burzlaff, and Zimmermann (1982), and Gubler (1982).

For nearly all space groups (exceptions are only the space-group types $Im\bar{3}m$ and $Ia\bar{3}d$) $N_A(G)$ is a proper supergroup of G. The elements of $N_A(G)$ leave G invariant as a whole but they may map a Wyckoff position $W_G(X_0)$ of G onto another Wyckoff position $W_G(X_0')$. The set of all Wyckoff positions on which $W_G(X_0)$ may be mapped by elements of $N_A(G)$ is called the Wyckoff set of X_0 or of $W_G(X_0)$.

In the above-mentioned example of a space group G = Pma2 the Wyckoff positions 2c m 1/4,y,z and 4d 1 x,y,z form a Wyckoff set each, whereas the Wyckoff positions 2a 2 0,0,z and 2b 2 0,1/2,z belong to the same Wyckoff set. Wyckoff positions with different types of site symmetry necessarily belong to different Wyckoff sets, those with site symmetries of the same type may (Positions 4e, 4f, 4g, 4h of a space group C222) or may not (Positions 4e, 4i of the same space group) belong to the same Wyckoff set. The Wyckoff sets are listed in IT 1983, p. 824 ff.

A Wyckoff set is designated by the letters of its Wyckoff positions, e.g. 2a-d, 4e-h, 4i,j, 4k, and 8l for a space group C222.

Different affine mappings of the space group G_1 onto G_2 may map a Wyckoff position W_{G_1} of G_1 onto different Wyckoff positions W_{G_2} of G_2. Therefore, the assignement of the Wyckoff position is not unique. Wyckoff sets, however, are mapped onto each other uniquely. In an analogous way as space groups are classified into space-group types, Wyckoff sets are classified into <u>types of Wyckoff sets</u> (cf., e.g. IT 1983, p. 725 f). To the 219+11=230 affine space-group types belong 1128+29 types of Wyckoff sets.

Example. The 12 Wyckoff positions of a space group C222 belong to the 5 Wyckoff sets 2a-d, 4e-h, 4i,j, 4k, and 8l respectively. Correspondingly there are 5 types of Wyckoff sets for the space-group type C222. Their symbols are the same as those for the Wyckoff sets, see above.

2.3. Eigensymmetry of a crystallographic orbit; crystallographic orbit type

Let us consider different crystallographic orbits of the same space group. Obviously the orbits of different Wyckoff positions may display different shapes, as in the example of the CaF_2 structure, space group $Fm\bar{3}m$, where the calcium atoms form an fcc point lattice and the fluorine atoms form a simple cubic point lattice (cf. section 6.2).

Moreover, the crystallographic orbits of the same Wyckoff position will differ in their appearance and their intrinsic symmetries, as the next example shows. In space group $P4_1$ we consider the following crystallographic orbits of the Wyckoff position 4a: $0,0,z_0$ is a tetragonal P point lattice with $c'=1/4c$, $1/2,0,z_0$ is a tetragonal I point lattice with $c'=1/2c$, and $1/10,1/5,z_0$ is not a point lattice but a set of tetragonal screws around lines parallel to the c-axis. Also the symmetries are different. The three crystallographic orbits "as such", i.e. disengaged from the original space group $P4_1$, display the symmetries P4/mmm, I4/mmm, and $P4_1$ respectively. The concept of a "crystallographic

orbit as such" is the basis of the following considerations.

Let $O_G(X_0)$ be the crystallographic orbit generated
from the point X_0 by the symmetry operations of the
space group G. The set $O_G(X_0)$ is a periodic set of
points which may be considered separately from the
generating space group G. The symmetry of this set
of points, i.e. the set of all motions leaving the
crystallographic orbit invariant as a whole, is
called the eigensymmetry space group E of the crystallographic orbit. Clearly, E is a space group and
E≥G holds, i.e., E may be a proper supergroup of the
generating space group G. If E=G, the crystallographic orbit is called a characteristic crystallographic orbit, otherwise, if E>G, it is called a
non-characteristic crystallographic orbit with
respect to G. Necessarily any crystallographic orbit
is always characteristic in the eigensymmetry space
group E. As a special case, $O_G(X_0)$ is called an
extraordinary orbit or an extraorbit if in E there
are additional translations not contained in G.

The following cases may be distinguished, if E>G
holds. They are displayed in Figure 1.

-G and E belong to the same crystal class. Then $G \overset{c}{<} E$
 is a class equivalent subgroup of E.

-G and E have the same translations. Then $G \overset{t}{<} E$ is a
 translation equivalent subgroup of E.

-G is a general subgroup of E, $G \overset{g}{<} E$, i.e., G is a
 proper subgroup of E, but neither class equivalent
 nor translation equivalent.

In this monograph it is assumed throughout that the
symmetry of the lattice is that necessary for the
space group G, such that there is no accidental
higher lattice symmetry. Moreover, accidental ratios
of the lattice parameters are excluded.

For example, in space group Pmmm, the crystallographic orbit 0,0,0 would have tetragonal or even
cubic symmetry if accidentally a=b (or b=c or c=a)
or a=b=c holds. Similarly the crystallographic
orbit 0,1/2,0 in space group P4/mmm would have cubic

symmetry if $c=a\sqrt{2}/2$. Both cases are not taken into account here in order to confine the amount of data.

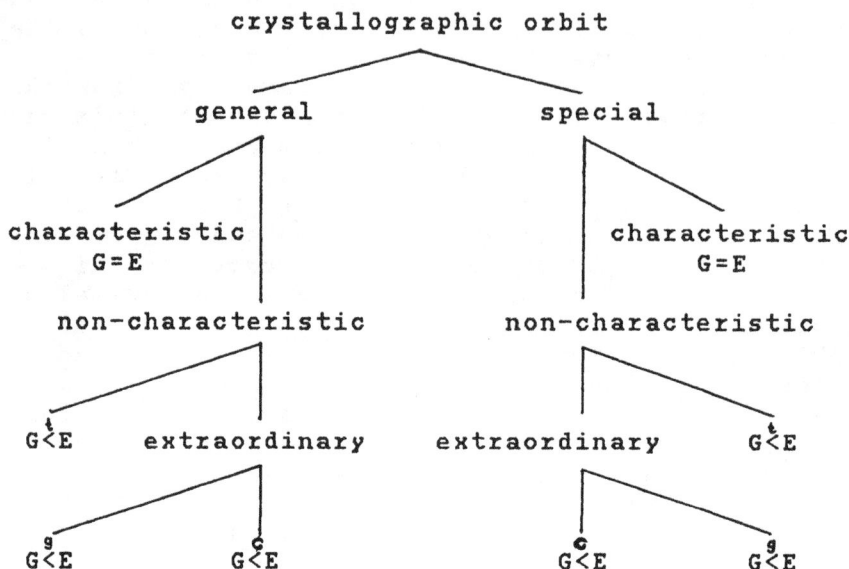

```
                    crystallographic orbit
                   /                     \
            general                       special
           /      |                      /       |
  characteristic  |          characteristic      |
      G=E         |              G=E             |
                  |                              |
        non-characteristic          non-characteristic
            /         |                   |          \
          G<E    extraordinary      extraordinary    G<E
          /    |                   |          \
        G<E   G<E                G<E          G<E
```

Fig. 1 The different kinds of crystallographic orbits

A classification of all crystallographic orbits by Wyckoff positions or Wyckoff sets is not possible, as a crystallographic orbit may have been generated by different space groups. For example, in the idealized perovskite structure (cf. section 6.4), space group $Pm\bar{3}m$, the partial structure of the oxygen atoms belongs to the Wyckoff position 3c 4/mmm or 3d 4/mmm, depending on the origin choice. It may be generated, however, also in other space groups from different Wyckoff positions, as shown in Table 1.

Crystallographic orbits can be classified, however, by using the eigensymmetry E. If E is the eigensym-metry of the crystallographic orbit $O_G(X_0)$ and if $S_0(X_0)$ is the site-symmetry group of X_0 in E, then any crystallographic orbit $O_{G'}(X_0')$ with eigensymmetry E' and site-symmetry group $S_0'(X_0')$ in E' belongs to the same orbit type as $O_G(X_0)$, if there is an affine mapping transforming E onto E' and simultaneously $S_0(X_0)$ onto $S_0'(X_0')$. In other words, two

Table 1. The occurence of orbit type Pm3̄m(3c,d) in
        cubic space groups

P23	1/2,0,0		3	d	222	
	0,1/2,1/2		3	c	222	
F23	0,1/4,1/4	ex	24	g	2	x,1/4,1/4
	1/4,0,0	ex	24	f	2	x,0,0
I23	1/4,1/4,0	ex	24	f	1	x,y,z
Pm3̄	1/2,0,0		3	d	mmm	
	0,1/2,1/2		3	c	mmm	
Pn3̄	1/4,0,0	ex	24	h	1	x,y,z
Fm3̄	1/4,0,0	ex	24	e	mm2	x,0,0
	0,1/4,1/4		24	d	2/m	
Im3̄	0,1/4,1/4	ex	24	g	m	0,y,z
Pa3̄	0,1/4,0	ex	24	d	1	x,y,z
	1/4,1/4,0	ex	24	d	1	x,y,z
Ia3̄	0,0,1/4	ex	24	d	2	x,0,1/4
	1/4,0,1/4	ex	24	d	2	x,0,1/4
P432	1/2,0,0		3	d	422	
	0,1/2,1/2		3	c	422	
P4$_2$32	0,1/4,1/4	ex	24	m	1	x,y,z
F432	1/4,0,0	ex	24	e	4	x,0,0
	0,1/4,1/4		24	d	222	
I432	0,1/4,1/4	ex	24	h	2	0,y,y
P4̄3m	1/2,0,0		3	d	4̄2m	
	0,1/2,1/2		3	c	4̄2m	
F4̄3m	0,1/4,1/4	ex	24	g	mm2	x,1/4,1/4
	1/4,0,0	ex	24	f	mm2	x,0,0
I4̄3m	1/4,1/4,0	ex	24	g	m	x,x,z
P4̄3n	1/4,1/4,0	ex	24	i	1	x,y,z
F4̄3c	1/4,0,0		24	d	4̄	
	1/4,1/4,0		24	c	4̄	
Pm3̄m	1/2,0,0		3	d	4/mmm	
	0,1/2,1/2		3	c	4/mmm	
Pn3̄n	1/4,0,0	ex	24	h	2	1/4,y,y
Pm3̄n	0,1/4,1/4	ex	24	k	m	0,y,z
Pn3̄m	0,0,1/4	ex	24	k	m	x,x,z
Fm3̄m	1/4,0,0	ex	24	e	4mm	x,0,0
	0,1/4,1/4		24	d	mmm	
Fm3̄c	0,1/4,1/4		24	d	4/m	
	1/4,0,0		24	c	4̄m2	
Fd3̄c	1/8,0,0	ex	192	h	1	x,y,z
Im3̄m	0,1/4,1/4	ex	24	h	mm2	0,y,y

Explanation. Columns 1-3: space group, representing
point, Wyckoff position. An "ex" indicates that the
Wyckoff position contains other crystallographic
orbits too. In these cases a representing manifold
is given in column 4.

orbits belong to the same orbit type if they belong to the same type of Wyckoff sets under the type of E, i.e. if they have the same symbol in the eigensymmetry space group, e.g. Pm$\bar{3}$m (3c,d).

2.4. Vector lattices and point lattices

The word "lattice" has been used in different contexts by scientists. In IT 1983, p. 714 it is the name for the set of all translation vectors of an ideal crystal structure, whereas earlier it has been used synonymously for crystal structure, as in "the NaCl lattice". Here we follow IT 1983. With the aid of "lattice" as set of vectors, point lattices may be defined in the following way. Given a point X_0 and a vector lattice of vectors $\vec{t}_j$. The set of all points X_j with $\overline{X_0 X_j} = \vec{t}_j$ is called the point lattice belonging to X_0. Analogously, a crystallographic orbit is called a point lattice if all of its points are translationally equivalent under some space group G'. The space group G' is not necessarily the generating space group G of the crystallographic orbit. The point lattice is called a real point lattice if its points are equivalent under the translations of the generating space group G. Otherwise it is called an apparent point lattice. For example, in space group P$\bar{1}$ the crystallographic orbit generated from the point 0,0,0 is a real point lattice, whereas the crystallographic orbit generated from the point 1/4,0,0 is an apparent point lattice. In a real point lattice not only the points but also their surroundings are translationally equivalent. In an apparent point lattice only the points but not their surroundings have this property.

Nowacki (1975) has shown that in 3-dimensional space groups, except those of types P4$_1$32 and P4$_3$32, at least one real or apparent point lattice exists.

3. Main Table of non-characteristic crystallographic orbits

3.0. Short explanation of the Main Table

Headline: No. and standard Hermann-Mauguin symbol of the space group according to IT 1983.

Setting and origin: Unique axis b in monoclinic space groups, hexagonal axes in trigonal space groups with rhombohedral lattice, origin choice 2 (in a centre of inversion).

Superlattices: List of superlattices (SL) occuring in the space group. SL number from Table 2. Symbol for the lattice type referred to the basis of the space group, and conventional basis $\vec{a}'$, $\vec{b}'$, $\vec{c}'$ of the superlattice in terms of the conventional basis of the space group (for a more detailed explanation cf. chapter 4). Density d: index of the original lattice in the superlattice. Additional generators (add. gen.): g_1, g_2, g_3 are short symbols for those vectors $g_1\vec{a}$ + $g_2\vec{b}$ + $g_3\vec{c}$ which generate, together with the basis of the space group, the superlattice.

In each space group the list of non-characteristic orbits is ordered according to the Wyckoff positions of the space group.

First column: multiplicity, Wyckoff letter, and site symmetry of the Wyckoff position of the space group.

Within the Wyckoff position the data are ordered according to the SL number, starting with the orbits of SL 0, the lattice of the space group.

Second column: SL number of the superlattice.

The following data are printed in one or two columns, each consisting of the orbits concerned (in accordance with the first entry of the corresponding data of IT 1983), followed by a conventional symbol of the eigensymmetry space group and the multiplicity and Wyckoff letters of the appropriate Wyckoff set (for the term Wyckoff set, cf. section 2.2.). The space-group symbol of the Eigensymmetry space group reflects the relation to the original space group and is, thus, not always the standard symbol, e.g. Abmm or Bmcm may be found instead of Cmma.

3.1. Triclinic system

Space group No. 1 P1

Superlattices

SL 0 P(1,1,1)   density 1 add.gen.

Wyckoff letter   non-characteristic crystallographic orbits

   1 a 1   SL 0 x,y,z   P$\bar{1}$ (1a-h)

Space group No. 2 P$\bar{1}$

Superlattices

SL 1 P(1/2,1,1)   density 2 add.gen.   1/2,0,0
SL 2 P(1,1/2,1)   density 2 add.gen.   0,1/2,0
SL 3 P(1,1,1/2)   density 2 add.gen.   0,0,1/2
SL 4 A(1,1,1)   density 2 add.gen.   0,1/2,1/2
SL 5 B(1,1,1)   density 2 add.gen.   1/2,0,1/2
SL 6 C(1,1,1)   density 2 add.gen.   1/2,1/2,0
SL 7 I(1,1,1)   density 2 add.gen.   1/2,1/2,1/2

Wyckoff letter   non-characteristic crystallographic orbits

2 i 1	SL 1	1/4,1/2,1/2	P$\bar{1}$ (1a-h)	1/4,1/2,0	P$\bar{1}$ (1a-h)
		1/4,0,1/2	P$\bar{1}$ (1a-h)	1/4,0,0	P$\bar{1}$ (1a-h)
	SL 2	1/2,1/4,1/2	P$\bar{1}$ (1a-h)	1/2,1/4,0	P$\bar{1}$ (1a-h)
		0,1/4,1/2	P$\bar{1}$ (1a-h)	0,1/4,0	P$\bar{1}$ (1a-h)
	SL 3	1/2,1/2,1/4	P$\bar{1}$ (1a-h)	1/2,0,1/4	P$\bar{1}$ (1a-h)
		0,1/2,1/4	P$\bar{1}$ (1a-h)	0,0,1/4	P$\bar{1}$ (1a-h)
	SL 4	1/2,3/4,1/4	P$\bar{1}$ (1a-h)	1/2,1/4,1/4	P$\bar{1}$ (1a-h)
		0,3/4,1/4	P$\bar{1}$ (1a-h)	0,1/4,1/4	P$\bar{1}$ (1a-h)
	SL 5	3/4,1/2,1/4	P$\bar{1}$ (1a-h)	3/4,0,1/4	P$\bar{1}$ (1a-h)
		1/4,1/2,1/4	P$\bar{1}$ (1a-h)	1/4,0,1/4	P$\bar{1}$ (1a-h)
	SL 6	1/4,3/4,1/2	P$\bar{1}$ (1a-h)	1/4,3/4,0	P$\bar{1}$ (1a-h)
		1/4,1/4,1/2	P$\bar{1}$ (1a-h)	1/4,1/4,0	P$\bar{1}$ (1a-h)
	SL 7	3/4,3/4,1/4	P$\bar{1}$ (1a-h)	3/4,1/4,1/4	P$\bar{1}$ (1a-h)
		1/4,3/4,1/4	P$\bar{1}$ (1a-h)	1/4,1/4,1/4	P$\bar{1}$ (1a-h)

3.2. Monoclinic system

Space group No. 3 P2

Superlattices

```
SL  0  P(1,1,1)      density  1  add.gen.
SL  1  P(1/2,1,1)    density  2  add.gen.  1/2,0,0
SL  3  P(1,1,1/2)    density  2  add.gen.  0,0,1/2
SL  5  B(1,1,1)      density  2  add.gen.  1/2,0,1/2
```

Wyckoff letter   non-characteristic crystallographic orbits

```
2 e 1     SL 0   x,y,z         P2/m (2m,n)
          SL 1   1/4,y,1/2     P2/m (1a-h)     1/4,y,0       P2/m (1a-h)
          SL 3   1/2,y,1/4     P2/m (1a-h)     0,y,1/4       P2/m (1a-h)
          SL 5   1/4,y,3/4     P2/m (1a-h)     1/4,y,1/4     P2/m (1a-h)
  1 d 2   SL 0   1/2,y,1/2     P2/m (1a-h)
  1 c 2   SL 0   1/2,y,0       P2/m (1a-h)
  1 b 2   SL 0   0,y,1/2       P2/m (1a-h)
  1 a 2   SL 0   0,y,0         P2/m (1a-h)
```

Space group No. 4 P2₁

Superlattices

```
SL  0  P(1,1,1)      density  1  add.gen.
SL  2  P(1,1/2,1)    density  2  add.gen.  0,1/2,0
SL  4  A(1,1,1)      density  2  add.gen.  0,1/2,1/2
SL  6  C(1,1,1)      density  2  add.gen.  1/2,1/2,0
SL  7  I(1,1,1)      density  2  add.gen.  1/2,1/2,1/2
```

Wyckoff letter   non-characteristic crystallographic orbits

```
2 a 1     SL 0   x,y,z         P2₁/m (2e)
          SL 2   1/2,y,1/2     P2/m (1a-h)     1/2,y,0       P2/m (1a-h)
                 0,y,1/2       P2/m (1a-h)     0,y,0         P2/m (1a-h)
          SL 4   1/2,y,1/4     A2/m (2a-d)     0,y,1/4       A2/m (2a-d)
          SL 6   1/4,y,1/2     C2/m (2a-d)     1/4,y,0       C2/m (2a-d)
          SL 7   1/4,y,3/4     I2/m (2a-d)     1/4,y,1/4     I2/m (2a-d)
```

Space group No. 5 C2

Superlattices

SL 0 C(1,1,1)   density 1 add.gen.
SL 1 P(1/2,1/2,1)   density 2 add.gen.   1/2,0,0
SL 2 C(1,1,1/2)   density 2 add.gen.   0,0,1/2
SL 3 F(1,1,1)   density 2 add.gen.   0,1/2,1/2

Wyckoff letter   non-characteristic crystallographic orbits

  4 c 1   SL 0 x,y,z   C2/m (4i)
             SL 1 1/4,y,1/2   P2/m (1a-h)   1/4,y,0   P2/m (1a-h)
             SL 2 0,y,1/4   C2/m (2a-d)
             SL 3 1/4,y,1/4   C2/m (2a-d)
  2 b 2   SL 0 0,y,1/2   C2/m (2a-d)
  2 a 2   SL 0 0,y,0   C2/m (2a-d)

Space group No. 6 Pm

Superlattices

SL 0 P(1,1,1)   density 1 add.gen.
SL 2 P(1,1/2,1)   density 2 add.gen.   0,1/2,0

Wyckoff letter   non-characteristic crystallographic orbits

  2 c 1   SL 0 x,y,z   P2/m (2i-1)
             SL 2 x,1/4,z   P2/m (1a-h)
  1 b m   SL 0 x,1/2,z   P2/m (1a-h)
  1 a m   SL 0 x,0,z   P2/m (1a-h)

Space group No. 7 Pc

Superlattices

```
SL  0  P(1,1,1)          density  1  add.gen.
SL  3  P(1,1,1/2)        density  2  add.gen.  0,0,1/2
SL  4  A(1,1,1)          density  2  add.gen.  0,1/2,1/2
```

Wyckoff letter   non-characteristic crystallographic orbits

```
   2 a 1    SL 0  x,y,z      P2/c (2e,f)
            SL 3  x,1/2,z    P2/m (1a-h)    x,0,z      P2/m (1a-h)
            SL 4  x,1/4,z    A2/m (2a-d)
```

Space group No. 8 Cm

Superlattices

```
SL  0  C(1,1,1)          density  1  add.gen.
SL  1  P(1/2,1/2,1)      density  2  add.gen.  1/2,0,0
```

Wyckoff letter   non-characteristic crystallographic orbits

```
   4 b 1    SL 0  x,y,z      C2/m (4g,h)
            SL 1  x,1/4,z    P2/m (1a-h)
   2 a m    SL 0  x,0,z      C2/m (2a-d)
```

Space group No. 9 Cc

Superlattices

```
SL  0  C(1,1,1)        density  1  add.gen.
SL  2  C(1,1,1/2)      density  2  add.gen.   0,0,1/2
SL  3  F(1,1,1)        density  2  add.gen.   0,1/2,1/2
```

Wyckoff letter   non-characteristic crystallographic orbits

```
4 a 1     SL 0   x,y,z        C2/c (4e)
          SL 2   x,0,z        C2/m (2a-d)
          SL 3   x,1/4,z      C2/m (2a-d)
```

Space group No. 10   P2/m

Superlattices

```
SL   1  P(1/2,1,1)     density  2  add.gen.   1/2,0,0
SL   2  P(1,1/2,1)     density  2  add.gen.   0,1/2,0
SL   3  P(1,1,1/2)     density  2  add.gen.   0,0,1/2
SL   5  B(1,1,1)       density  2  add.gen.   1/2,0,1/2
SL   9  P(1/2,1/2,1)   density  4  add.gen.   1/2,0,0   0,1/2,0
SL  16  P(1,1/2,1/2)   density  4  add.gen.   0,1/2,0   0,0,1/2
SL  18  B(1,1/2,1)     density  4  add.gen.   0,1/2,0   1/2,0,1/2
```

Wyckoff letter   non-characteristic crystallographic orbits

```
4 o 1     SL 1   1/4,y,1/2      P2/m (2i-1)      1/4,y,0        P2/m (2i-1)
          SL 2   x,1/4,z        P2/m (2m,n)
          SL 3   1/2,y,1/4      P2/m (2i-1)      0,y,1/4        P2/m (2i-1)
          SL 5   3/4,y,1/4      P2/m (2i-1)      1/4,y,1/4      P2/m (2i-1)
          SL 9   1/4,1/4,1/2    P2/m (1a-h)      1/4,1/4,0      P2/m (1a-h)
          SL16   1/2,1/4,1/4    P2/m (1a-h)      0,1/4,1/4      P2/m (1a-h)
          SL18   3/4,1/4,1/4    P2/m (1a-h)      1/4,1/4,1/4    P2/m (1a-h)
2 n m     SL 1   1/4,1/2,1/2    P2/m (1a-h)      1/4,1/2,0      P2/m (1a-h)
          SL 3   1/2,1/2,1/4    P2/m (1a-h)      0,1/2,1/4      P2/m (1a-h)
          SL 5   3/4,1/2,1/4    P2/m (1a-h)      1/4,1/2,1/4    P2/m (1a-h)
2 m m     SL 1   1/4,0,1/2      P2/m (1a-h)      1/4,0,0        P2/m (1a-h)
          SL 3   1/2,0,1/4      P2/m (1a-h)      0,0,1/4        P2/m (1a-h)
          SL 5   3/4,0,1/4      P2/m (1a-h)      1/4,0,1/4      P2/m (1a-h)
2 l 2     SL 2   1/2,1/4,1/2    P2/m (1a-h)
2 k 2     SL 2   0,1/4,1/2      P2/m (1a-h)
2 j 2     SL 2   1/2,1/4,0      P2/m (1a-h)
2 i 2     SL 2   0,1/4,0        P2/m (1a-h)
```

Space group No. 11 P2$_1$/m

Superlattices

SL	2	P(1,1/2,1)	density 2	add.gen.	0,1/2,0	
SL	4	A(1,1,1)	density 2	add.gen.	0,1/2,1/2	
SL	6	C(1,1,1)	density 2	add.gen.	1/2,1/2,0	
SL	7	I(1,1,1)	density 2	add.gen.	1/2,1/2,1/2	
SL	9	P(1/2,1/2,1)	density 4	add.gen.	1/2,0,0	0,1/2,0
SL	15	P(1,1/4,1)	density 4	add.gen.	0,1/4,0	
SL	16	P(1,1/2,1/2)	density 4	add.gen.	0,1/2,0	0,0,1/2
SL	18	B(1,1/2,1)	density 4	add.gen.	0,1/2,0	1/2,0,1/2

Wyckoff letter   non-characteristic crystallographic orbits

Wyckoff letter	SL					
4 f 1	SL 2	x,0,z	P2/m (2m,n)	1/2,y,1/2	P2/m (2i-l)	
		1/2,y,0	P2/m (2i-l)	0,y,1/2	P2/m (2i-l)	
		0,y,0	P2/m (2i-l)			
	SL 4	1/2,y,1/4	A2/m (4g,h)	0,y,1/4	A2/m (4g,h)	
	SL 6	1/4,y,1/2	C2/m (4g,h)	1/4,y,0	C2/m (4g,h)	
	SL 7	3/4,y,1/4	I2/m (4g,h)	1/4,y,1/4	I2/m (4g,h)	
	SL 9	1/4,0,1/2	P2/m (1a-h)	1/4,0,0	P2/m (1a-h)	
	SL15	1/2,1/8,1/2	P2/m (1a-h)	1/2,1/8,0	P2/m (1a-h)	
		0,1/8,1/2	P2/m (1a-h)	0,1/8,0	P2/m (1a-h)	
	SL16	1/2,0,1/4	P2/m (1a-h)	0,0,1/4	P2/m (1a-h)	
	SL18	3/4,0,1/4	P2/m (1a-h)	1/4,0,1/4	P2/m (1a-h)	
2 e m	SL 2	1/2,1/4,1/2	P2/m (1a-h)	1/2,1/4,0	P2/m (1a-h)	
		0,1/4,1/2	P2/m (1a-h)	0,1/4,0	P2/m (1a-h)	
	SL 4	1/2,1/4,3/4	A2/m (2a-d)	1/2,1/4,1/4	A2/m (2a-d)	
		0,1/4,3/4	A2/m (2a-d)	0,1/4,1/4	A2/m (2a-d)	
	SL 6	3/4,1/4,1/2	C2/m (2a-d)	1/4,1/4,1/2	C2/m (2a-d)	
		3/4,1/4,0	C2/m (2a-d)	1/4,1/4,0	C2/m (2a-d)	
	SL 7	1/4,1/4,3/4	I2/m (2a-d)	3/4,1/4,1/4	I2/m (2a-d)	
		3/4,1/4,3/4	I2/m (2a-d)	1/4,1/4,1/4	I2/m (2a-d)	
2 d $\bar{1}$	SL 2	1/2,0,1/2	P2/m (1a-h)			
2 c $\bar{1}$	SL 2	0,0,1/2	P2/m (1a-h)			
2 b $\bar{1}$	SL 2	1/2,0,0	P2/m (1a-h)			
2 a $\bar{1}$	SL 2	0,0,0	P2/m (1a-h)			

Space group   No. 12   C2/m

Superlattices

SL	1	P(1/2,1/2,1)	density	2	add.gen.	1/2,0,0	
SL	2	C(1,1,1/2)	density	2	add.gen.	0,0,1/2	
SL	3	F(1,1,1)	density	2	add.gen.	0,1/2,1/2	
SL	4	P(1/4,1/2,1)	density	4	add.gen.	1/4,0,0	
SL	5	P(1/2,1/4,1)	density	4	add.gen.	0,1/4,0	
SL	6	P(1/2,1/2,1/2)	density	4	add.gen.	1/2,0,0 0,0,1/2	
SL	8	B(1/2,1/2,1)	density	4	add.gen.	1/4,0,1/2	

Wyckoff letter   non-characteristic crystallographic orbits

Wyckoff	SL					
8 j 1	SL 1	x,1/4,z	P2/m (2m,n)	1/4,y,1/2	P2/m (2i-1)
		1/4,y,0	P2/m (2i-1)		
	SL 2	0,y,1/4	C2/m (4g,h)		
	SL 3	1/4,y,1/4	C2/m (4g,h)		
	SL 4	1/8,1/4,1/2	P2/m (1a-h)	1/8,1/4,0	P2/m (1a-h)
	SL 5	1/4,1/8,1/2	P2/m (1a-h)	1/4,1/8,0	P2/m (1a-h)
	SL 6	1/4,1/4,1/4	P2/m (1a-h)	0,1/4,1/4	P2/m (1a-h)
	SL 8	1/8,1/4,1/4	P2/m (1a-h)	3/8,1/4,1/4	P2/m (1a-h)
4 i m	SL 1	1/4,0,1/2	P2/m (1a-h)	1/4,0,0	P2/m (1a-h)
	SL 2	0,0,1/4	C2/m (2a-d)	1/2,0,1/4	C2/m (2a-d)
	SL 3	3/4,0,1/4	C2/m (2a-d)	1/4,0,1/4	C2/m (2a-d)
4 h 2	SL 1	0,1/4,1/2	P2/m (1a-h)		
4 g 2	SL 1	0,1/4,0	P2/m (1a-h)		
4 f $\bar{1}$	SL 1	1/4,1/4,1/2	P2/m (1a-h)		
4 e $\bar{1}$	SL 1	1/4,1/4,0	P2/m (1a-h)		

Space group No. 13 P2/c

Superlattices

```
SL  1  P(1/2,1,1)      density  2  add.gen.  1/2,0,0
SL  3  P(1,1,1/2)      density  2  add.gen.  0,0,1/2
SL  4  A(1,1,1)        density  2  add.gen.  0,1/2,1/2
SL  5  B(1,1,1)        density  2  add.gen.  1/2,0,1/2
SL 10  P(1/2,1,1/2)    density  4  add.gen.  1/2,0,0   0,0,1/2
SL 11  A(1/2,1,1)      density  4  add.gen.  1/2,0,0   0,1/2,1/2
SL 16  P(1,1/2,1/2)    density  4  add.gen.  0,1/2,0   0,0,1/2
SL 21  P(1,1,1/4)      density  4  add.gen.  0,0,1/4
SL 23  B(1,1,1/2)      density  4  add.gen.  1/2,0,1/4
SL 26  F(1,1,1)        density  4  add.gen.  0,1/2,1/2   1/2,0,1/2
```

Wyckoff letter   non-characteristic crystallographic orbits

```
  4 g 1    SL 1   1/4,y,1/4      P2/c (2e,f)
           SL 3   x,1/2,z        P2/m (2m,n)     x,0,z         P2/m (2m,n)
                  1/2,y,0        P2/m (2i-1)     0,y,0         P2/m (2₁-1)
           SL 4   x,1/4,z        A2/m (4i)
           SL 5   1/4,y,0        P2/c (2e,f)
           SL10   1/4,1/2,0      P2/m (1a-h)     1/4,0,0       P2/m (1a-h)
                  1/4,1/2,1/4    P2/m (1a-h)     1/4,0,1/4     P2/m (1a-h)
           SL11   1/4,1/4,3/4    A2/m (2a-d)     1/4,1/4,1/4   A2/m (2a-d)
           SL16   1/2,1/4,0      P2/m (1a-h)     0,1/4,0       P2/m (1a-h)
           SL21   1/2,1/2,1/8    P2/m (1a-h)     1/2,0,1/8     P2/m (1a-h)
                  0,1/2,1/8      P2/m (1a-h)     0,0,1/8       P2/m (1a-h)
           SL23   1/4,1/2,3/8    P2/m (1a-h)     1/4,0,3/8     P2/m (1a-h)
                  1/4,1/2,1/8    P2/m (1a-h)     1/4,0,1/8     P2/m (1a-h)
           SL26   1/4,1/4,1/2    C2/m (2a-d)     1/4,1/4,0     C2/m (2a-d)
  2 f 2    SL 3   1/2,1/2,1/4    P2/m (1a-h)     1/2,0,1/4     P2/m (1a-h)
           SL 4   1/2,3/4,1/4    A2/m (2a-d)     1/2,1/4,1/4   A2/m (2a-d)
  2 e 2    SL 3   0,1/2,1/4      P2/m (1a-h)     0,0,1/4       P2/m (1a-h)
           SL 4   0,3/4,1/4      A2/m (2a-d)     0,1/4,1/4     A2/m (2a-d)
  2 d 1̄    SL 3   1/2,0,0        P2/m (1a-h)
  2 c 1̄    SL 3   0,1/2,0        P2/m (1a-h)
  2 b 1̄    SL 3   1/2,1/2,0      P2/m (1a-h)
  2 a 1̄    SL 3   0,0,0          P2/m (1a-h)
```

Space group No. 14 P2₁/c

Superlattices

```
SL   2   P(1,1/2,1)      density  2  add.gen.  0,1/2,0
SL   3   P(1,1,1/2)      density  2  add.gen.  0,0,1/2
SL   4   A(1,1,1)        density  2  add.gen.  0,1/2,1/2
SL   6   C(1,1,1)        density  2  add.gen.  1/2,1/2,0
SL   7   I(1,1,1)        density  2  add.gen.  1/2,1/2,1/2
SL  11   A(1/2,1,1)      density  4  add.gen.  1/2,0,0   0,1/2,1/2
SL  16   P(1,1/2,1/2)    density  4  add.gen.  0,1/2,0   0,0,1/2
SL  17   A(1,1/2,1)      density  4  add.gen.  0,1/4,1/2
SI  22   A(1,1,1/2)      density  4  add.gen.  0,1/2,1/4
SL  24   C(1,1,1/2)      density  4  add.gen.  0,0,1/2   1/2,1/2,0
SL  25   I(1,1,1/2)      density  4  add.gen.  1/2,1/2,1/4
SL  26   F(1,1,1)        density  4  add.gen.  0,1/2,1/2  1/2,0,1/2
```

Wyckoff letter   non-characteristic crystallographic orbits

Wyckoff letter	SL					
4 e 1	SL 2	1/2,y,1/4	P2/c (2e,f)	0,y,1/4	P2/c (2e,f)
	SL 3	x,1/4,z	P2₁/m (2e)		
	SL 4	x,0,z	A2/m (4i)	1/2,y,0	A2/m (4g,h)
		0,y,0	A2/m (4g,h)		
	SL 6	1/4,y,1/4	C2/c (4e)		
	SL 7	1/4,y,0	I2/a (4e)		
	SL11	1/4,0,1/2	A2/m (2a-d)	1/4,0,0	A2/m (2a-d)
	SL16	1/2,1/4,0	P2/m (1a-h)	0,1/4,0	P2/m (1a-h)
		1/2,0,1/4	P2/m (1a-h)	0,0,1/4	P2/m (1a-h)
		1/2,1/4,1/4	P2/m (1a-h)	0,1/4,1/4	P2/m (1a-h)
	SL17	1/2,3/8,1/4	A2/m (2a-d)	0,3/8,1/4	A2/m (2a-d)
		1/2,1/8,1/4	A2/m (2a-d)	0,1/8,1/4	A2/m (2a-d)
	SL22	1/2,1/4,3/8	A2/m (2a-d)	0,1/4,3/8	A2/m (2a-d)
		1/2,1/4,1/8	A2/m (2a-d)	0,1/4,1/8	A2/m (2a-d)
	SL24	3/4,1/4,0	C2/m (2a-d)	1/4,1/4,0	C2/m (2a-d)
		3/4,1/4,1/4	C2/m (2a-d)	1/4,1/4,1/4	C2/m (2a-d)
	SL25	1/4,1/4,3/8	I2/m (2a-d)	3/4,1/4,3/8	I2/m (2a-d)
		3/4,1/4,1/8	I2/m (2a-d)	1/4,1/4,1/8	I2/m (2a-d)
	SL26	1/4,1/2,1/4	C2/m (2a-d)	1/4,0,1/4	C2/m (2a-d)
2 d 1̄	SL 4	1/2,0,1/2	A2/m (2a-d)		
2 c 1̄	SL 4	0,0,1/2	A2/m (2a-d)		
2 b 1̄	SL 4	1/2,0,0	A2/m (2a-d)		
2 a 1̄	SL 4	0,0,0	A2/m (2a-d)		

Space group No. 15 C2/c

Superlattices

```
SL  1   P(1/2,1/2,1)     density  2   add.gen.   1/2,0,0
SL  2   C(1,1,1/2)·      density  2   add.gen.   0,0,1/2
SL  3   F(1,1,1)         density  2   add.gen.   0,1/2,1/2
SL  6   P(1/2,1/2,1/2)   density  4   add.gen.   1/2,0,0   0,0,1/2
SL  7   A(1/2,1/2,1)     density  4   add.gen.   0,1/4,1/2
SL 11   C(1,1,1/4)       density  4   add.gen.   0,0,1/4
SL 12   F(1,1,1/2)       density  4   add.gen.   0,1/2,1/4
SL 13   C(1,1,1/2*)      density  4   add.gen.   1/4,0,1/4
SL 14   F(1,1,1*)        density  4   add.gen.   1/4,1/2,1/4
```

Wyckoff letter   non-characteristic crystallographic orbits

Wyckoff letter		SL		orbit	symmetry	orbit	symmetry
8	f 1	SL 1	1/4,y,1/4	P2/c (2e,f)			
		SL 2	x,0,z	C2/m (4i)	0,y,0	C2/m (4g,h)
		SL 3	x,1/4,z	C2/m (4i)	1/4,y,0	C2/m (4g,h)
		SL 6	0,1/4,0	P2/m (1a-h)	1/4,0,0	P2/m (1a-h)
			1/4,1/4,1/4	P2/m (1a-h)	1/4,0,1/4	P2/m (1a-h)
		SL 7	1/4,3/8,1/4	A2/m (2a-d)	1/4,1/8,1/4	A2/m (2a-d)
		SL11	0,1/2,1/8	C2/m (2a-d)	0,0,1/8	C2/m (2a-d)
		SL12	1/4,1/2,1/8	C2/m (2a-d)	1/4,0,1/8	C2/m (2a-d)
		SL13	1/8,1/4,3/8	C2/m (2a-d)	3/8,1/4,1/8	C2/m (2a-d)
		SL14	3/8,1/4,3/8	C2/m (2a-d)	1/8,1/4,1/8	C2/m (2a-d)
4	e 2	SL 2	0,1/2,1/4	C2/m (2a-d)	0,0,1/4	C2/m (2a-d)
		SL 3	0,3/4,1/4	C2/m (2a-d)	0,1/4,1/4	C2/m (2a-d)
4	d $\bar{1}$	SL 3	1/4,1/4,1/2	C2/m (2a-d)		
4	c $\bar{1}$	SL 3	1/4,1/4,0	C2/m (2a-d)		
4	b $\bar{1}$	SL 2	0,1/2,0	C2/m (2a-d)		
4	a $\bar{1}$	SL 2	0,0,0	C2/m (2a-d)		

3.3. Orthorhombic system

Space group No. 16 P222

Superlattices

```
SL  0  P(1,1,1)        density 1  add.gen.
SL  1  P(1/2,1,1)      density 2  add.gen.  1/2,0,0
SL  2  P(1,1/2,1)      density 2  add.gen.  0,1/2,0
SL  3  P(1,1,1/2)      density 2  add.gen.  0,0,1/2
SL  4  A(1,1,1)        density 2  add.gen.  0,1/2,1/2
SL  5  B(1,1,1)        density 2  add.gen.  1/2,0,1/2
SL  6  C(1,1,1)        density 2  add.gen.  1/2,1/2,0
SL  9  P(1/2,1/2,1)    density 4  add.gen.  1/2,0,0  0,1/2,0
SL 10  P(1/2,1,1/2)    density 4  add.gen.  1/2,0,0  0,0,1/2
SL 16  P(1,1/2,1/2)    density 4  add.gen.  0,1/2,0  0,0,1/2
SL 26  F(1,1,1)        density 4  add.gen.  0,1/2,1/2  1/2,0,1/2
```

Wyckoff letter   non-characteristic crystallographic orbits

```
  4 u 1    SL 0   x,y,1/2        Pmmm (4u-z)    x,y,0        Pmmm (4u-z)
                  x,y,1/4        Pccm (4q)      x,1/2,z      Pmmm (4u-z)
                  x,0,z          Pmmm (4u-z)    x,1/4,z      Pbmb (4q)
                  1/2,y,z        Pmmm (4u-z)    0,y,z        Pmmm (4u-z)
                  1/4,y,z        Pmaa (4q)
           SL 1   1/4,y,1/2      Pmmm (2i-t)    1/4,y,0      Pmmm (2i-t)
                  1/4,1/2,z      Pmmm (2i-t)    1/4,0,z      Pmmm (2i-t)
           SL 2   x,1/4,1/2      Pmmm (2i-t)    x,1/4,0      Pmmm (2i-t)
                  1/2,1/4,z      Pmmm (2i-t)    0,1/4,z      Pmmm (2i-t)
           SL 3   x,1/2,1/4      Pmmm (2i-t)    x,0,1/4      Pmmm (2i-t)
                  1/2,y,1/4      Pmmm (2i-t)    0,y,1/4      Pmmm (2i-t)
           SL 4   x,1/4,1/4      Abmm (4g)
           SL 5   1/4,y,1/4      Bmcm (4g)
           SL 6   1/4,1/4,z      Cmma (4g)
           SL 9   1/4,1/4,1/2    Pmmm (1a-h)    1/4,1/4,0    Pmmm (1a-h)
           SL10   1/4,1/2,1/4    Pmmm (1a-h)    1/4,0,1/4    Pmmm (1a-h)
           SL16   1/2,1/4,1/4    Pmmm (1a-h)    0,1/4,1/4    Pmmm (1a-h)
           SL26   3/4,1/4,1/4    Fmmm (4a,b)    1/4,1/4,1/4  Fmmm (4a,b)
  2 t 2    SL 0   1/2,1/2,z      Pmmm (2i-t)
           SL 3   1/2,1/2,1/4    Pmmm (1a-h)
  2 s 2    SL 0   0,1/2,z        Pmmm (2i-t)
           SL 3   0,1/2,1/4      Pmmm (1a-h)
  2 r 2    SL 0   1/2,0,z        Pmmm (2i-t)
           SL 3   1/2,0,1/4      Pmmm (1a-h)
  2 q 2    SL 0   0,0,z          Pmmm (2i-t)
           SL 3   0,0,1/4        Pmmm (1a-h)
  2 p 2    SL 0   1/2,y,1/2      Pmmm (2i-t)
           SL 2   1/2,1/4,1/2    Pmmm (1a-h)
  2 o 2    SL 0   1/2,y,0        Pmmm (2i-t)
           SL 2   1/2,1/4,0      Pmmm (1a-h)
  2 n 2    SL 0   0,y,1/2        Pmmm (2i-t)
           SL 2   0,1/4,1/2      Pmmm (1a-h)
  2 m 2    SL 0   0,y,0          Pmmm (2i-t)
           SL 2   0,1/4,0        Pmmm (1a-h)
```

(continued on next page)

Space group No. 16 P222   (continued)

Wyckoff letter   non-characteristic crystallographic orbits

```
2 l 2      SL 0   x,1/2,1/2      Pmmm (2i-t)
           SL 1   1/4,1/2,1/2    Pmmm (1a-h)
2 k 2      SL 0   x,1/2,0        Pmmm (2i-t)
           SL 1   1/4,1/2,0      Pmmm (1a-h)
2 j 2      SL 0   x,0,1/2        Pmmm (2i-t)
           SL 1   1/4,0,1/2      Pmmm (1a-h)
2 i 2      SL 0   x,0,0          Pmmm (2i-t)
           SL 1   1/4,0,0        Pmmm (1a-h)
1 h 222    SL 0   1/2,1/2,1/2    Pmmm (1a-h)
1 g 222    SL 0   0,1/2,1/2      Pmmm (1a-h)
1 f 222    SL 0   1/2,0,1/2      Pmmm (1a-h)
1 e 222    SL 0   1/2,1/2,0      Pmmm (1a-h)
1 d 222    SL 0   0,0,1/2        Pmmm (1a-h)
1 c 222    SL 0   0,1/2,0        Pmmm (1a-h)
1 b 222    SL 0   1/2,0,0        Pmmm (1a-h)
1 a 222    SL 0   0,0,0          Pmmm (1a-h)
```

Space group No. 17 $P222_1$

Superlattices

SL 0	P(1,1,1)	density	1	add.gen.		
SL 1	P(1/2,1,1)	density	2	add.gen.	1/2,0,0	
SL 2	P(1,1/2,1)	density	2	add.gen.	0,1/2,0	
SL 3	P(1,1,1/2)	density	2	add.gen.	0,0,1/2	
SL 4	A(1,1,1)	density	2	add.gen.	0,1/2,1/2	
SL 5	B(1,1,1)	density	2	add.gen.	1/2,0,1/2	
SL 7	I(1,1,1)	density	2	add.gen.	1/2,1/2,1/2	
SL 10	P(1/2,1,1/2)	density	4	add.gen.	1/2,0,0	0,0,1/2
SL 11	A(1/2,1,1)	density	4	add.gen.	1/2,0,0	0,1/2,1/2
SL 16	P(1,1/2,1/2)	density	4	add.gen.	0,1/2,0	0,0,1/2
SL 18	B(1,1/2,1)	density	4	add.gen.	0,1/2,0	1/2,0,1/2
SL 21	P(1,1,1/4)	density	4	add.gen.	0,0,1/4	

Wyckoff letter		non-characteristic crystallographic orbits			
4 e 1	SL 0	x,y,0	Pcmm (4k)	x,1/2,z	Pcmm (4i,j)
		x,0,z	Pcmm (4i,j)	x,y,1/4	Pmcm (4k)
		1/2,y,z	Pmcm (4i,j)	0,y,z	Pmcm (4i,j)
		1/4,y,z	Pmna (4h)	x,1/4,z	Pnmb (4h)
	SL 1	1/4,y,1/4	Pmcm (2e,f)		
	SL 2	x,1/4,0	Pcmm (2e,f)		
	SL 3	x,1/2,1/4	Pmmm (2i-t)	x,0,1/4	Pmmm (2i-t)
		1/2,1/2,z	Pmmm (2i-t)	1/2,0,z	Pmmm (2i-t)
		0,1/2,z	Pmmm (2i-t)	0,0,z	Pmmm (2i-t)
		1/2,y,0	Pmmm (2i-t)	0,y,0	Pmmm (2i-t)
	SL 4	x,1/4,3/4	Ammm (4k,l)	x,1/4,1/4	Ammm (4k,l)
		1/2,1/4,z	Ammm (4g-j)	0,1/4,z	Ammm (4g-j)
	SL 5	1/4,1/2,z	Bmmm (4g-j)	1/4,0,z	Bmmm (4g-j)
		1/4,y,1/2	Bmmm (4k,l)	1/4,y,0	Bmmm (4k,l)
	SL 7	1/4,1/4,z	Imma (4e)		
	SL10	1/4,1/2,1/4	Pmmm (1a-h)	1/4,0,1/4	Pmmm (1a-h)
	SL11	1/4,1/4,3/4	Ammm (2a-d)	1/4,1/4,1/4	Ammm (2a-d)
	SL16	1/2,1/4,0	Pmmm (1a-h)	0,1/4,0	Pmmm (1a-h)
	SL18	1/4,1/4,1/2	Bmmm (2a-d)	1/4,1/4,0	Bmmm (2a-d)
	SL21	1/2,1/2,1/8	Pmmm (1a-h)	1/2,0,1/8	Pmmm (1a-h)
		0,1/2,1/8	Pmmm (1a-h)	0,0,1/8	Pmmm (1a-h)
2 d 2	SL 0	1/2,y,1/4	Pmcm (2e,f)		
	SL 3	1/2,1/2,1/4	Pmmm (1a-h)	1/2,0,1/4	Pmmm (1a-h)
	SL 4	1/2,3/4,1/4	Ammm (2a-d)	1/2,1/4,1/4	Ammm (2a-d)
2 c 2	SL 0	0,y,1/4	Pmcm (2e,f)		
	SL 3	0,1/2,1/4	Pmmm (1a-h)	0,0,1/4	Pmmm (1a-h)
	SL 4	0,3/4,1/4	Ammm (2a-d)	0,1/4,1/4	Ammm (2a-d)
2 b 2	SL 0	x,1/2,0	Pcmm (2e,f)		
	SL 3	1/2,1/2,0	Pmmm (1a-h)	0,1/2,0	Pmmm (1a-h)
	SL 5	3/4,1/2,0	Bmmm (2a-d)	1/4,1/2,0	Bmmm (2a-d)
2 a 2	SL 0	x,0,0	Pcmm (2e,f)		
	SL 3	1/2,0,0	Pmmm (1a-h)	0,0,0	Pmmm (1a-h)
	SL 5	3/4,0,0	Bmmm (2a-d)	1/4,0,0	Bmmm (2a-d)

Space group No. 18   P2₁2₁2

Superlattices

```
SL  0  P(1,1,1)        density 1  add.gen.
SL  1  P(1/2,1,1)      density 2  add.gen.  1/2,0,0
SL  2  P(1,1/2,1)      density 2  add.gen.  0,1/2,0
SL  4  A(1,1,1)        density 2  add.gen.  0,1/2,1/2
SL  5  B(1,1,1)        density 2  add.gen.  1/2,0,1/2
SL  6  C(1,1,1)        density 2  add.gen.  1/2,1/2,0
SL  7  I(1,1,1)        density 2  add.gen.  1/2,1/2,1/2
SI  9  P(1/2,1/2,1)    density 4  add.gen.  1/2,0,0  0,1/2,0
SL 11  A(1/2,1,1)      density 4  add.gen.  1/2,0,0  0,1/2,1/2
SL 13  C(1/2,1,1)      density 4  add.gen.  1/4,1/2,0
SL 18  B(1,1/2,1)      density 4  add.gen.  0,1/2,0  1/2,0,1/2
SL 19  C(1,1/2,1)      density 4  add.gen.  1/2,1/4,0
SL 26  F(1,1,1)        density 4  add.gen.  0,1/2,1/2  1/2,0,1/2
```

Wyckoff letter   non-characteristic crystallographic orbits

Wyckoff letter	SL					
4 c 1	SL 0	x,y,1/2	Pbam (4g,h)	x,y,0	Pbam (4g,h)
		1/4,y,z	Pmab (4d)	x,1/4,z	Pbma (4d)
		x,y,1/4	Pnnm (4g)	x,0,z	Pmmn (4e,f)
		0,y,z	Pmmn (4e,f)		
	SL 1	1/4,0,z	Pmmb (2e,f)	x,1/4,1/2	Pbmm (2e,f)
		x,1/4,0	Pbmm (2e,f)		
	SL 2	0,1/4,z	Pmma (2e,f)	1/4,y,1/2	Pmam (2e,f)
		1/4,y,0	Pmam (2e,f)		
	SL 4	1/4,y,1/4	Amam (4c)		
	SL 5	x,1/4,1/4	Bbmm (4c)		
	SL 6	1/4,1/4,z	Cmma (4g)	x,0,1/2	Cmmm (4g-j)
		x,0,0	Cmmm (4g-j)	0,y,1/2	Cmmm (4g-j)
		0,y,0	Cmmm (4g-j)		
	SL 7	x,0,3/4	Imm (4e-j)	x,0,1/4	Imm (4e-j)
		0,y,3/4	Imm (4e-j)	0,y,1/4	Imm (4e-j)
	SL 9	1/4,0,1/2	Pmmm (1a-h)	1/4,0,0	Pmmm (1a-h)
		0,1/4,1/2	Pmmm (1a-h)	0,1/4,0	Pmmm (1a-h)
		1/4,1/4,1/2	Pmmm (1a-h)	1/4,1/4,0	Pmmm (1a-h)
	SL11	1/4,0,3/4	Ammm (2a-d)	1/4,0,1/4	Ammm (2a-d)
	SL13	3/8,1/4,1/2	Cmmm (2a-d)	3/8,1/4,0	Cmmm (2a-d)
		1/8,1/4,1/2	Cmmm (2a-d)	1/8,1/4,0	Cmmm (2a-d)
	SL18	0,1/4,3/4	Bmmm (2a-d)	0,1/4,1/4	Bmmm (2a-d)
	SL19	1/4,3/8,1/2	Cmmm (2a-d)	1/4,3/8,0	Cmmm (2a-d)
		1/4,1/8,1/2	Cmmm (2a-d)	1/4,1/8,0	Cmmm (2a-d)
	SL26	1/4,1/4,3/4	Fmmm (4a,b)	1/4,1/4,1/4	Fmmm (4a,b)
2 b 2	SL 0	0,1/2,z	Pmmn (2a,b)		
	SL 6	0,1/2,1/2	Cmmm (2a-d)	0,1/2,0	Cmmm (2a-d)
	SL 7	0,1/2,3/4	Imm (2a-d)	0,1/2,1/4	Imm (2a-d)
2 a 2	SL 0	0,0,z	Pmmn (2a,b)		
	SL 6	0,0,1/2	Cmmm (2a-d)	0,0,0	Cmmm (2a-d)
	SL 7	0,0,3/4	Imm (2a-d)	0,0,1/4	Imm (2a-d)

Space group No. 19 P2₁2₁2₁

Superlattices

```
SL  0   P(1,1,1)        density  1   add.gen.
SL  1   P(1/2,1,1)      density  2   add.gen.   1/2,0,0
SL  2   P(1,1/2,1)      density  2   add.gen.   0,1/2,0
SL  3   P(1,1,1/2)      density  2   add.gen.   0,0,1/2
SL  4   A(1,1,1)        density  2   add.gen.   0,1/2,1/2
SL  5   B(1,1,1)        density  2   add.gen.   1/2,0,1/2
SL  6   C(1,1,1)        density  2   add.gen.   1/2,1/2,0
SL  7   I(1,1,1)        density  2   add.gen.   1/2,1/2,1/2
SL 11   A(1/2,1,1)      density  4   add.gen.   1/2,0,0   0,1/2,1/2
SL 14   I(1/2,1,1)      density  4   add.gen.   1/4,1/2,1/2
SL 18   B(1,1/2,1)      density  4   add.gen.   0,1/2,0   1/2,0,1/2
SL 20   I(1,1/2,1)      density  4   add.gen.   1/2,1/4,1/2
SL 24   C(1,1,1/2)      density  4   add.gen.   0,0,1/2   1/2,1/2,0
SL 25   I(1,1,1/2)      density  4   add.gen.   1/2,1/2,1/4
SL 26   F(1,1,1)        density  4   add.gen.   0,1/2,1/2  1/2,0,1/2
```

Wyckoff letter   non-characteristic crystallographic orbits

```
    4 a 1   SL 0   x,y,0          Pnam (4c)       x,y,1/4         Pbnm (4c)
                   x,0,z          Pcmn (4c)       x,1/4,z         Pnma (4c)
                   0,y,z          Pmnb (4c)       1/4,y,z         Pmcn (4c)
            SL 1   x,1/4,0        Pnmm (2a,b)     x,1/4,1/2       Pnmm (2a,b)
            SL 2   0,y,1/4        Pmnm (2a,b)     1/2,y,1/4       Pmnm (2a,b)
            SL 3   1/4,0,z        Pmmn (2a,b)     1/4,1/2,z       Pmmn (2a,b)
            SL 4   0,y,0          Amam (4c)       1/4,1/4,z       Amma (4c)
            SL 5   0,0,z          Bmmb (4c)       x,1/4,1/4       Bbmm (4c)
            SL 6   x,0,0          Ccmm (4c)       1/4,y,1/4       Cmcm (4c)
            SL 7   x,0,1/4        Icmm (4e)       1/4,y,0         Imam (4e)
                   0,1/4,z        Immb (4e)
            SL11   0,1/4,0        Ammm (2a-d)     0,1/4,1/2       Ammm (2a-d)
                   1/4,1/4,0      Ammm (2a-d)     1/4,1/4,1/2     Ammm (2a-d)
            SL14   1/8,1/4,0      Immm (2a-d)     1/8,1/4,1/2     Immm (2a-d)
                   3/8,1/4,0      Immm (2a-d)     3/8,1/4,1/2     Immm (2a-d)
            SL18   0,0,1/4        Bmmm (2a-d)     1/2,0,1/4       Bmmm (2a-d)
                   0,1/4,1/4      Bmnm (2a-d)     1/2,1/4,1/4     Bmmm (2a-d)
            SL20   0,1/8,1/4      Immm (2a-d)     1/2,1/8,1/4     Immm (2a-d)
                   0,3/8,1/4      Immm (2a-d)     1/2,3/8,1/4     Immm (2a-d)
            SL24   1/4,0,0        Cmmm (2a-d)     1/4,1/2,0       Cmmm (2a-d)
                   1/4,0,1/4      Cmmm (2a-d)     1/4,1/2,1/4     Cmmm (2a-d)
            SL25   1/4,0,1/8      Immm (2a-d)     1/4,1/2,1/8     Immm (2a-d)
                   1/4,0,3/8      Immm (2a-d)     1/4,1/2,3/8     Immm (2a-d)
            SL26   0,0,0          Fmmm (4a,b)     0,1/2,0         Fmmm (4a,b)
                   1/4,1/4,1/4    Fmmm (4a,b)     1/4,1/4,3/4     Fmmm (4a,b)
```

Space group No. 20 C222$_1$

Superlattices

```
SL  0  C(1,1,1)         density  1  add.gen.
SL  1  P(1/2,1/2,1)     density  2  add.gen.  1/2,0,0
SL  2  C(1,1,1/2)       density  2  add.gen.  0,0,1/2
SL  3  F(1,1,1)    `     density  2  add.gen.  0,1/2,1/2
SL  6  P(1/2,1/2,1/2)   density  4  add.gen.  1/2,0,0   0,0,1/2
SL  7  A(1/2,1/2,1)     density  4  add.gen.  0,1/4,1/2
SL  8  B(1/2,1/2,1)     density  4  add.gen.  1/4,0,1/2
SL 11  C(1,1,1/4)       density  4  add.gen.  0,0,1/4
SL 12  F(1,1,1/2)       density  4  add.gen.  0,1/2,1/4
```

Wyckoff letter   non-characteristic crystallographic orbits

8 c 1	SL 0	x,y,1/4	Cmcm (8g)	0,y,z	Cmcm (8f)
		x,y,0	Ccmm (8g)	x,0,z	Ccmm (8f)
		1/4,y,z	Cmca (8f)	x,1/4,z	Ccmb (8f)
	SL 1	x,1/4,0	Pcmm (2e,f)	1/4,y,1/4	Pmcm (2e,f)
	SL 2	1/4,1/4,z	Cmma (4g)	x,0,1/4	Cmmm (4g-j)
		0,1/2,z	Cmmm (4k,l)	0,0,z	Cmmm (4k,l)
		0,y,0	Cmmm (4g-j)		
	SL 3	1/4,0,z	Fmmm (8g-i)	1/4,y,0	Fmmm (8g-i)
		x,1/4,1/4	Fmmm (8g-i)	0,1/4,z	Fmmm (8g-i)
	SL 6	1/4,1/4,0	Pmmm (1a-h)	0,1/4,0	Pmmm (1a-h)
		1/4,1/4,1/4	Pmmm (1a-h)	1/4,0,1/4	Pmmm (1a-h)
	SL 7	1/4,3/8,1/4	Ammm (2a-d)	1/4,1/8,1/4	Ammm (2a-d)
	SL 8	3/8,1/4,0	Bmmm (2a-d)	1/8,1/4,0	Bmmm (2a-d)
	SL11	0,1/2,1/8	Cmmm (2a-d)	0,0,1/8	Cmmm (2a-d)
	SL12	1/4,1/4,3/8	Fmmm (4a,b)	1/4,1/4,1/8	Fmmm (4a,b)
4 b 2	SL 0	0,y,1/4	Cmcm (4c)		
	SL 2	0,1/2,1/4	Cmmm (2a-d)	0,0,1/4	Cmmm (2a-d)
	SL 3	0,3/4,1/4	Fmmm (4a,b)	0,1/4,1/4	Fmmm (4a,b)
4 a 2	SL 0	x,0,0	Ccmm (4c)		
	SL 2	1/2,0,0	Cmmm (2a-d)	0,0,0	Cmmm (2a-d)
	SL 3	3/4,0,0	Fmmm (4a,b)	1/4,0,0	Fmmm (4a,b)

Space group No. 21 C222

Superlattices

```
SL  0  C(1,1,1)        density  1  add.gen.
SL  1  P(1/2,1/2,1)    density  2  add.gen.  1/2,0,0
SL  2  C(1,1,1/2)      density  2  add.gen.  0,0,1/2
SL  3  F(1,1,1)        density  2  add.gen.  0,1/2,1/2
SL  4  P(1/4,1/2,1)    density  4  add.gen.  1/4,0,0
SL  5  P(1/2,1/4,1)    density  4  add.gen.  0,1/4,0
SL  6  P(1/2,1/2,1/2)  density  4  add.gen.  1/2,0,0   0,0,1/2
```

Wyckoff letter   non-characteristic crystallographic orbits

```
8 l 1     SL 0   x,y,1/2       Cmmm (8p,q)    x,y,0        Cmmm (8p,q)
                 x,0,z         Cmmm (8n,o)    0,y,z        Cmmm (8n,o)
                 x,y,1/4       Cccm (8l)      x,1/4,z      Cmma (8m,n)
                 1/4,y,z       Cmma (8m,n)
          SL 1   x,1/4,1/2     Pmmm (2i-t)    x,1/4,0      Pmmm (2i-t)
                 0,1/4,z       Pmmm (2i-t)    1/4,y,1/2    Pmmm (2i-t)
                 1/4,y,0       Pmmm (2i-t)    1/4,0,z      Pmmm (2i-t)
          SL 2   x,0,1/4       Cmmm (4g-j)    0,y,1/4      Cmmm (4g-j)
          SL 3   1/4,y,1/4     Fmmm (8g-i)    x,1/4,1/4    Fmmm (8g-i)
          SL 4   1/8,1/4,1/2   Pmmm (1a-h)    1/8,1/4,0    Pmmm (1a-h)
          SL 5   1/4,1/8,1/2   Pmmm (1a-h)    1/4,1/8,0    Pmmm (1a-h)
          SL 6   0,1/4,1/4     Pmmm (1a-h)    1/4,0,1/4    Pmmm (1a-h)
4 k 2     SL 0   1/4,1/4,z     Cmma (4g)
          SL 1   1/4,1/4,1/2   Pmmm (1a-h)    1/4,1/4,0    Pmmm (1a-h)
          SL 3   1/4,1/4,3/4   Fmmm (4a,b)    1/4,1/4,1/4  Fmmm (4a,b)
4 j 2     SL 0   0,1/2,z       Cmmm (4k,l)
          SL 2   0,1/2,1/4     Cmmm (2a-d)
4 i 2     SL 0   0,0,z         Cmmm (4k,l)
          SL 2   0,0,1/4       Cmmm (2a-d)
4 h 2     SL 0   0,y,1/2       Cmmm (4g-j)
          SL 1   0,1/4,1/2     Pmmm (1a-h)
4 g 2     SL 0   0,y,0         Cmmm (4g-j)
          SL 1   0,1/4,0       Pmmm (1a-h)
4 f 2     SL 0   x,0,1/2       Cmmm (4g-j)
          SL 1   1/4,0,1/2     Pmmm (1a-h)
4 e 2     SL 0   x,0,0         Cmmm (4g-j)
          SL 1   1/4,0,0       Pmmm (1a-h)
2 d 222   SL 0   0,0,1/2       Cmmm (2a-d)
2 c 222   SL 0   1/2,0,1/2     Cmmm (2a-d)
2 b 222   SL 0   0,1/2,0       Cmmm (2a-d)
2 a 222   SL 0   0,0,0         Cmmm (2a-d)
```

Space group No. 22 F222

Superlattices

```
SL  0  F(1,1,1)        density  1  add.gen.
SL  1  P(1/2,1/2,1/2)  density  2  add.gen.  1/2,0,0
SL  2  P(1/4,1/2,1/2)  density  4  add.gen.  1/4,0,0
SL  3  P(1/2,1/4,1/2)  density  4  add.gen.  0,1/4,0
SL  4  P(1/2,1/2,1/4)  density  4  add.gen.  0,0,1/4
```

Wyckoff letter   non-characteristic crystallographic orbits

```
16 k 1   SL 0  x,y,0          Fmmm (16m-o)   x,y,1/4        Fmmm (16m-o)
               x,0,z          Fmmm (16m-o)   x,1/4,z        Fmmm (16m-o)
               0,y,z          Fmmm (16m-o)   1/4,y,z        Fmmm (16m-o)
               1/8,1/8,1/8    Fddd (16c,d)   3/8,3/8,3/8    Fddd (16c,d)
               5/8,5/8,5/8    Fddd (16c,d)   7/8,7/8,7/8    Fddd (16c,d)
         SL 1  x,0,1/4        Pmmm (2i-t)    x,1/4,0        Pmmm (2i-t)
               0,y,1/4        Pmmm (2i-t)    1/4,y,0        Pmmm (2i-t)
               0,1/4,z        Pmmm (2i-t)    1/4,0,z        Pmmm (2i-t)
         SL 2  1/8,0,1/4      Pmmm (1a-h)    1/8,1/4,0      Pmmm (1a-h)
         SL 3  0,1/8,1/4      Pmmm (1a-h)    1/4,1/8,0      Pmmm (1a-h)
         SL 4  0,1/4,1/8      Pmmm (1a-h)    1/4,0,1/8      Pmmm (1a-h)
 8 j 2   SL 0  x,1/4,1/4      Fmmm (8g-i)
         SL 1  0,1/4,1/4      Pmmm (1a-h)
 8 i 1   SL 0  1/4,y,1/4      Fmmm (8g-i)
         SL 1  1/4,0,1/4      Pmmm (1a-h)
 8 h 1   SL 0  1/4,1/4,z      Fmmm (8g-i)
         SL 1  1/4,1/4,0      Pmmm (1a-h)
 8 g 1   SL 0  0,0,z          Fmmm (8g-i)
         SL 1  0,0,1/4        Pmmm (1a-h)
 8 f 1   SL 0  0,y,0          Fmmm (8g-i)
         SL 1  0,1/4,0        Pmmm (1a-h)
 8 e 2   SL 0  x,0,0          Fmmm (8g-i)
         SL 1  1/4,0,0        Pmmm (1a-h)
 4 d 222 SL 0  1/4,1/4,3/4    Fmmm (4a,b)
 4 c 222 SL 0  1/4,1/4,1/4    Fmmm (4a,b)
 4 b 222 SL 0  0,0,1/2        Fmmm (4a,b)
 4 a 222 SL 0  0,0,0          Fmmm (4a,b)
```

Space group No. 23 I222

Superlattices

```
SL  0   I(1,1,1)          density  1  add.gen.
SL  1   A(1/2,1,1)        density  2  add.gen.  1/2,0,0
SL  2   B(1,1/2,1)        density  2  add.gen.  0,1/2,0
SL  3   C(1,1,1/2)        density  2  add.gen.  0,0,1/2
SL  4   P(1/2,1/2,1/2)    density  4  add.gen.  1/2,0,0  0,1/2,0
SL  6   F(1/2,1,1)        density  4  add.gen.  1/4,0,1/2
S1  8   F(1,1/2,1)        density  4  add.gen.  0,1/4,1/2
SL 10   F(1,1,1/2)        density  4  add.gen.  0,1/2,1/4
```

Wyckoff letter   non-characteristic crystallographic orbits

```
8 k 1    SL 0   x,y,0          Immm (8l-n)    x,0,z          Immm (8l-n)
                0,y,z          Immm (8l-n)    x,y,1/4        Ibam (8j)
                1/4,y,z        Imcb (8j)      x,1/4,z        Icma (8j)
         SL 1   x,1/4,1/4      Abmm (4g)      1/4,y,0        Ammm (4g-j)
                1/4,0,z        Ammm (4g-j)
         SL 2   1/4,y,1/4      Bmcm (4g)      x,1/4,0        Bmmm (4g-j)
                0,1/4,z        Bmmm (4g-j)
         SL 3   1/4,1/4,z      Cmma (4g)      x,0,1/4        Cmmm (4g-j)
              . 0,y,1/4        Cmmm (4g-j)
         SL 4   1/4,1/4,1/4    Pmmm (1a-h)    0,1/4,1/4      Pmmm (1a-h)
                1/4,1/4,0      Pmmm (1a-h)    1/4,0,1/4      Pmmm (1a-h)
         SL 6   3/8,1/4,1/4    Fmmm (4a,b)    1/8,1/4,1/4    Fmmm (4a,b)
         SL 8   1/4,3/8,1/4    Fmmm (4a,b)    1/4,1/8,1/4    Fmmm (4a,b)
         SL10   1/4,1/4,3/8    Fmmm (4a,b)    1/4,1/4,1/8    Fmmm (4a,b)
4 j 2    SL 0   0,1/2,z        Immm (4e-j)
         SL 3   0,1/2,1/4      Cmmm (2a-d)
4 i 2    SL 0   0,0,z          Immm (4e-j)
         SL 3   0,0,1/4        Cmmm (2a-d)
4 h 2    SL 0   1/2,y,0        Immm (4e-j)
         SL 2   1/2,1/4,0      Bmmm (2a-d)
4 g 2    SL 0   0,y,0          Immm (4e-j)
         SL 2   0,1/4,0        Bmmm (2a-d)
4 f 2    SL 0   x,0,1/2        Immm (4e-j)
         SL 1   1/4,0,1/2      Ammm (2a-d)
4 e 2    SL 0   x,0,0          Immm (4e-j)
         SL 1   1/4,0,0        Ammm (2a-d)
2 d 222  SL 0   0,1/2,0        Immm (2a-d)
2 c 222  SL 0   0,0,1/2        Immm (2a-d)
2 b 222  SL 0   1/2,0,0        Immm (2a-d)
2 a 222  SL 0   0,0,0          Immm (2a-d)
```

Space group   No. 24 $I2_1 2_1 2_1$

Superlattices

```
SL  0  I(1,1,1)          density  1  add.gen.
SL  1  A(1/2,1,1)        density  2  add.gen.  1/2,0,0
SL  2  B(1,1/2,1)        density  2  add.gen.  0,1/2,0
SL  3  C(1,1,1/2)        density  2  add.gen.  0,0,1/2
SL  4  P(1/2,1/2,1/2)    density  4  add.gen.  1/2,0,0   0,1/2,0
SL  5  A(1/4,1,1)        density  4  add.gen.  1/4,0,0
SL  7  B(1,1/4,1)        density  4  add.gen.  0,1/4,0
SL  9  C(1,1,1/4)        density  4  add.gen.  0,0,1/4
```

Wyckoff letter   non-characteristic crystallographic orbits

```
8 d 1    SL 0   x,1/4,z        Imma (8i,h)     0,y,z          Imma (8i,h)
                x,y,1/4        Ibmm (8i,h)     x,0,z          Ibmm (8i,h)
                1/4,y,z        Imcm (8i,h)     x,y,0          Imcm (8i,h)
         SL 1   0,y,0          Ammm (4g-j)     1/4,1/4,z      Ammm (4g-j)
                x,1/4,1/2      Ammm (4k,l)     x,1/4,0        Ammm (4k,l)
         SL 2   0,0,z          Bmmm (4g-j)     x,1/4,1/4      Bmmm (4g-j)
                1/2,y,1/4      Bmmm (4k,l)     0,y,1/4        Bmmm (4k,l)
         SL 3   x,0,0          Cmmm (4g-j)     1/4,y,1/4      Cmmm (4g-j)
                1/4,1/2,z      Cmmm (4k,l)     1/4,0,z        Cmmm (4k,l)
         SL 4   0,0,0          Pmmm (1a-h)     1/4,1/4,1/4    Pmmm (1a-h)
         SL 5   1/8,1/4,1/2    Ammm (2a-d)     1/8,1/4,0      Ammm (2a-d)
         SL 7   1/2,1/8,1/4    Bmmm (2a-d)     0,1/8,1/4      Bmmm (2a-d)
         SL 9   1/4,1/2,1/8    Cmmm (2a-d)     1/4,0,1/8      Cmmm (2a-d)
4 c 2    SL 0   0,1/4,z        Imma (4e)
         SL 1   0,1/4,1/2      Ammm (2a-d)     0,1/4,0        Ammm (2a-d)
         SL 2   0,1/4,3/4      Bmmm (2a-d)     0,1/4,1/4      Bmmm (2a-d)
4 b 2    SL 0   1/4,y,0        Imcm (4e)
         SL 1   1/4,3/4,0      Ammm (2a-d)     1/4,1/4,0      Ammm (2a-d)
         SL 3   1/4,1/2,0      Cmmm (2a-d)     1/4,0,0        Cmmm (2a-d)
4 a 2    SL 0   x,0,1/4        Ibmm (4e)
         SL 2   1/2,0,1/4      Bmmm (2a-d)     0,0,1/4        Bmmm (2a-d)
         SL 3   3/4,0,1/4      Cmmm (2a-d)     1/4,0,1/4      Cmmm (2a-d)
```

Space group No. 25 Pmm2

Superlattices

```
SL  0  P(1,1,1)        density  1  add.gen.
SL  1  P(1/2,1,1)      density  2  add.gen.  1/2,0,0
SL  2  P(1,1/2,1)      density  2  add.gen.  0,1/2,0
SL  9  P(1/2,1/2,1)    density  4  add.gen.  1/2,0,0   0,1/2,0
```

Wyckoff letter   non-characteristic crystallographic orbits

```
    4 i 1      SL 0   x,y,z          Pmmm (4u-z)
               SL 1   1/4,y,z        Pmmm (2i-t)
               SL 2   x,1/4,z        Pmmm (2i-t)
               SL 9   1/4,1/4,z      Pmmm (1a-h)
    2 h m      SL 0   1/2,y,z        Pmmm (2i-t)
               SL 2   1/2,1/4,z      Pmmm (1a-h)
    2 g m      SL 0   0,y,z          Pmmm (2i-t)
               SL 2   0,1/4,z        Pmmm (1a-h)
    2 f m      SL 0   x,1/2,z        Pmmm (2i-t)
               SL 1   1/4,1/2,z      Pmmm (1a-h)
    2 e m      SL 0   x,0,z          Pmmm (2i-t)
               SL 1   1/4,0,z        Pmmm (1a-h)
·   1 d mm2    SL 0   1/2,1/2,z      Pmmm (1a-h)
    1 c mm2    SL 0   1/2,0,z        Pmmm (1a-h)
    1 b mm2    SL 0   0,1/2,z        Pmmm (1a-h)
    1 a mm2    SL 0   0,0,z          Pmmm (1a-h)
```


Space group No. 26 Pmc2$_1$

Superlattices

```
SL  0  P(1,1,1)        density  1  add.ger.
SL  1  P(1/2,1,1)      density  2  add.gen.  1/2,0,0
SL  3  P(1,1,1/2)      density  2  add.gen.  0,0,1/2
SL  4  A(1,1,1)        density  2  add.gen.  0,1/2,1/2
SL 10  P(1/2,1,1/2)    density  4  add.gen.  1/2,0,0   0,0,1/2
SL 11  A(1/2,1,1)      density  4  add.gen.  1/2,0,0   0,1/2,1/2
```

Wyckoff letter   non-characteristic crystallographic orbits

```
    4 c 1      SL 0   x,y,z          Pmcm (4k)
               SL 1   1/4,y,z        Pmcm (2e,f)
               SL 3   x,1/2,z        Pmmm (2i-t)     x,0,z      Pmmm (2i-t)
               SL 4   x,1/4,z        Ammm (4k,l)
               SL10   1/4,1/2,z      Pmmm (1a-h)     1/4,0,z    Pmmm (1a-h)
               SL11   1/4,1/4,z      Ammm (2a-d)
    2 b m      SL 0   1/2,y,z        Pmcm (2e,f)
               SL 3   1/2,1/2,z      Pmmm (1a-h)     1/2,0,z    Pmmm (1a-h)
               SL 4   1/2,1/4,z      Ammm (2a-d)
    2 a m      SL 0   0,y,z          Pmcm (2e,f)
               SL 3   0,1/2,z        Pmmm (1a-h)     0,0,z      Pmmm (1a-h)
               SL 4   0,1/4,z        Ammm (2a-d)
```

Space group No. 27 Pcc2

Superlattices

```
SL  0  P(1,1,1)        density  1  add.gen.
SL  3  P(1,1,1/2)      density  2  add.gen.  0,0,1/2
SL  4  A(1,1,1)        density  2  add.gen.  0,1/2,1/2
SL  5  B(1,1,1)        density  2  add.gen.  1/2,0,1/2
SL 10  P(1/2,1,1/2)    density  4  add.gen.  1/2,0,0   0,0,1/2
SL 16  P(1,1/2,1/2)    density  4  add.gen.  0,1/2,0   0,0,1/2
SL 26  F(1,1,1)        density  4  add.gen.  0,1/2,1/2  1/2,0,1/2
```

Wyckoff letter   non-characteristic crystallographic orbits

```
   4 e 1    SL 0   x,y,z        Pccm (4q)
            SL 3   1/2,y,z      Pmmm (2i-t)    0,y,z      Pmmm (2i-t)
                   x,1/2,z      Pmmm (2i-t)    x,0,z      Pmmm (2i-t)
            SL 4   x,1/4,z      Acmm (4g)
            SL 5   1/4,y,z      Bmcm (4g)
            SL10   1/4,1/2,z    Pmmm (1a-h)    1/4,0,z    Pmmm (1a-h)
            SL16   1/2,1/4,z    Pmmm (1a-h)    0,1/4,z    Pmmm (1a-h)
            SL26   1/4,1/4,z    Fmmm (4a,b)
   2 d 2    SL 3   1/2,1/2,z    Pmmm (1a-h)
   2 c 2    SL 3   1/2,0,z      Pmmm (1a-h)
   2 b 2    SL 3   0,1/2,z      Pmmm (1a-h)
   2 a 2    SL 3   0,0,z        Pmmm (1a-h)
```

Space group No. 28 Pma2

Superlattices

```
SL  0  P(1,1,1)        density  1  add.gen.
SL  1  P(1/2,1,1)      density  2  add.gen.  1/2,0,0
SL  6  C(1,1,1)        density  2  add.gen.  1/2,1/2,0
SL  8  P(1/4,1,1)      density  4  add.gen.  1/4,0,0
SL  9  P(1/2,1/2,1)    density  4  add.gen.  1/2,0,0   0,1/2,0
```

Wyckoff letter   non-characteristic crystallographic orbits

```
   4 d 1    SL 0   x,y,z        Pmam (4i,j)
            SL 1   0,y,z        Pmmm (2i-t)    x,1/2,z    Pmmm (2i-t)
                   x,0,z        Pmmm (2i-t)
            SL 6   x,1/4,z      Cmmm (4g-j)
            SL 8   1/8,1/2,z    Pmmm (1a-h)    1/8,0,z    Pmmm (1a-h)
            SL 9   0,1/4,z      Pmmm (1a-h)
   2 c m    SL 0   1/4,y,z      Pmam (2e,f)
            SL 1   1/4,1/2,z    Pmmm (1a-h)    1/4,0,z    Pmmm (1a-h)
            SL 6   1/4,1/4,z    Cmmm (2a-d)    1/4,3/4,z  Cmmm (2a-d)
   2 b 2    SL 1   0,1/2,z      Pmmm (1a-h)
   2 a 2    SL 1   0,0,z        Pmmm (1a-h)
```

Space group No. 29 Pca2₁

Superlattices

```
SL  0  P(1,1,1)        density  1  add.gen.
SL  1  P(1/2,1,1)      density  2  add.gen.  1/2,0,0
SL  3  P(1,1,1/2)      density  2  add.gen.  0,0,1/2
SL  5  B(1,1,1)        density  2  add.gen.  1/2,0,1/2
SL  6  C(1,1,1)        density  2  add.gen.  1/2,1/2,0
SL 10  P(1/2,1,1/2)    density  4  add.gen.  1/2,0,0   0,0,1/2
SL 12  B(1/2,1,1)      density  4  add.gen.  1/4,0,1/2
SL 24  C(1,1,1/2)      density  4  add.gen.  0,0,1/2   1/2,1/2,0
SL 26  F(1,1,1)        density  4  add.gen.  0,1/2,1/2  1/2,0,1/2
```

Wyckoff letter   non-characteristic crystallographic orbits

```
   4 a 1    SL 0   x,y,z       Pcam (4d)
            SL 1   x,1/2,z     Pcmm (2e,f)      x,0,z        Pcmm (2e,f)
            SL 3   1/4,y,z     Pmam (2e,f)
            SL 5   0,y,z       Bmam (4g)
            SL 6   x,1/4,z     Ccmm (4c)
            SL10   1/4,1/2,z   Pmmm (1a-h)     1/4,0,z       Pmmm (1a-h)
                   0,1/2,z     Pmmm (1a-h)     0,0,z         Pmmm (1a-h)
            SL12   1/8,1/2,z   Bmmm (2a-d)     1/8,0,z       Bmmm (2a-d)
            SL24   1/4,3/4,z   Cmmm (2a-d)     1/4,1/4,z     Cmmm (2a-d)
            SL26   0,1/4,z     Fmmm (4a,b)
```


Space group No. 30 Pnc2

Superlattices

```
SL  0  P(1,1,1)        density  1  add.gen.
SL  3  P(1,1,1/2)      density  2  add.gen.  0,0,1/2
SL  4  A(1,1,1)        density  2  add.gen.  0,1/2,1/2
SL  7  I(1,1,1)        density  2  add.gen.  1/2,1/2,1/2
SL 11  A(1/2,1,1)      density  4  add.gen.  1/2,0,0   0,1/2,1/2
SL 16  P(1,1/2,1/2)    density  4  add.gen.  0,1/2,0   0,0,1/2
SL 24  C(1,1,1/2)      density  4  add.gen.  0,0,1/2   1/2,1/2,0
```

Wyckoff letter   non-characteristic crystallographic orbits

```
   4 c 1    SL 0   x,y,z       Pncm (4h)
            SL 3   x,1/4,z     Pbmm (2e,f)
            SL 4   1/2,y,z     Ammm (4g-j)      0,y,z        Ammm (4g-j)
                   x,0,z       Ammm (4k,l)
            SL 7   1/4,y,z     Imcm (4e)
            SL11   1/4,0,z     Ammm (2a-d)
            SL16   1/2,1/4,z   Pmmm (1a-h)     0,1/4,z       Pmmm (1a-h)
            SL24   3/4,1/4,z   Cmmm (2a-d)     1/4,1/4,z     Cmmm (2a-d)
   2 b 2    SL 4   1/2,0,z     Ammm (2a-d)
   2 a 2    SL 4   0,0,z       Ammm (2a-d)
```

Space group   No. 31   Pmn2₁

Superlattices

```
SL  0   P(1,1,1)          density  1   add.gen.
SL  1   P(1/2,1,1)        density  2   add.gen.   1/2,0,0
SL  5   B(1,1,1)          density  2   add.gen.   1/2,0,1/2
SL  7   I(1,1,1)          density  2   add.gen.   1/2,1/2,1/2
SL 10   P(1/2,1,1/2)      density  4   add.gen.   1/2,0,0   0,0,1/2
SI 11   A(1/2,1,1)        density  4   add.gen.   1/2,0,0   0,1/2,1/2
```

Wyckoff letter   non-characteristic crystallographic orbits

```
   4 b 1     SL 0   x,y,z       Pmnm (4e,f)
             SL 1   1/4,y,z     Pmcm (2e,f)
             SL 5   x,1/2,z     Bmmm (4g-j)     x,0,z        Bmmm (4g-j)
             SL 7   x,1/4,z     Immm (4e-j)
             SL10   1/4,1/2,z   Pmmm (1a-h)     1/4,0,z      Pmmm (1a-h)
             SL11   1/4,1/4,z   Ammm (2a-d)
   2 a m     SL 0   0,y,z       Pmnm (2a,b)
             SL 5   0,1/2,z     Bmmm (2a-d)     0,0,z        Bmmm (2a-d)
             SL 7   0,3/4,z     Immm (2a-d)     0,1/4,z      Immm (2a-d)
```


Space group   No. 32   Pba2

Superlattices

```
SL  0   P(1,1,1)          density  1   add.gen.
SL  1   P(1/2,1,1)        density  2   add.gen.   1/2,0,0
SL  2   P(1,1/2,1)        density  2'  add.gen.   0,1/2,0
SL  6   C(1,1,1)          density  2   add.gen.   1/2,1/2,0
SL  9   P(1/2,1/2,1)      density  4   add.gen.   1/2,0,0   0,1/2,0
SL 13   C(1/2,1,1)        density  4   add.gen.   1/4,1/2,0
SL 19   C(1,1/2,1)        density  4   add.gen.   1/2,1/4,0
```

Wyckoff letter   non-characteristic crystallographic orbits

```
   4 c 1     SL 0   x,y,z       Pbam (4g,h)
             SL 1   x,1/4,z     Pbmm (2e,f)
             SL 2   1/4,y,z     Pmam (2e,f)
             SL 6   x,0,z       Cmmm (4g-j)     0,y,z        Cmmm (4g-j)
             SL 9   1/4,0,z     Pmmm (1a-h)     0,1/4,z      Pmmm (1a-h)
                    1/4,1/4,z   Pmmm (1a-h)
             SL13   3/8,1/4,z   Cmmm (2a-d)     1/8,1/4,z    Cmmm (2a-d)
             SL19   1/4,3/8,z   Cmmm (2a-d)     1/4,1/8,z    Cmmm (2a-d)
   2 b 2     SL 6   0,1/2,z     Cmmm (2a-d)
   2 a 2     SL 6   0,0,z       Cmmm (2a-d)
```

Space group No. 33 Pna2₁

Superlattices

```
SL  0   P(1,1,1)        density  1   add.gen.
SL  1   P(1/2,1,1)      density  2   add.gen.  1/2,0,0
SL  4   A(1,1,1)        density  2   add.gen.  0,1/2,1/2
SL  6   C(1,1,1)        density  2   add.gen.  1/2,1/2,0
SL  7   I(1,1,1)        density  2   add.gen.  1/2,1/2,1/2
SL 11   A(1/2,1,1)      density  4   add.gen.  1/2,0,0   0,1/2,1/2
SL 14   I(1/2,1,1)      density  4   add.gen.  1/4,1/2,1/2
SL 24   C(1,1,1/2)      density  4   add.gen.  0,0,1/2   1/2,1/2,0
SL 26   F(1,1,1)        density  4   add.gen.  0,1/2,1/2   1/2,0,1/2
```

Wyckoff letter   non-characteristic crystallographic orbits

```
    4 a 1     SL 0   x,y,z        Pnam (4c)
              SL 1   x,1/4,z      Pnmm (2a,b)
              SL 4   1/4,y,z      Amam (4c)
              SL 6   x,0,z        Ccmm (4c)
              SL 7   0,y,z        Imam (4e)
              SL11   0,1/4,z      Ammm (2a-d)    1/4,1/4,z   Ammm (2a-d)
              SL14   3/8,1/4,z    Immm (2a-d)    1/8,1/4,z   Immm (2a-d)
              SL24   0,1/2,z      Cmmm (2a-d)    0,0,z       Cmmm (2a-d)
              SL26   1/4,0,z      Fmmm (4a,b)
```

Space group No. 34 Pnn2

Superlattices

```
SL  0   P(1,1,1)        density  1   add.gen.
SL  4   A(1,1,1)        density  2   add.gen.  0,1/2,1/2
SL  5   B(1,1,1)        density  2   add.gen.  1/2,0,1/2
SL  7   I(1,1,1)        density  2   add.gen.  1/2,1/2,1/2
SL 11   A(1/2,1,1)      density  4   add.gen.  1/2,0,0   0,1/2,1/2
SL 18   B(1,1/2,1)      density  4   add.gen.  0,1/2,0   1/2,0,1/2
SL 26   F(1,1,1)        density  4   add.gen.  0,1/2,1/2   1/2,0,1/2
```

Wyckoff letter   non-characteristic crystallographic orbits

```
    4 c 1     SL 0   x,y,z        Pnnm (4g)
              SL 4   1/4,y,z      Amam (4c)
              SL 5   x,1/4,z      Bbmm (4c)
              SL 7   0,y,z        Immm (4e-j)    x,0,z       Immm (4e-j)
              SL11   1/4,0,z      Ammm (2a-d)
              SL18   0,1/4,z      Bmmm (2a-d)
              SL26   1/4,1/4,z    Fmmm (4a,b)
    2 b 2     SL 7   0,1/2,z      Immm (2a-d)
    2 a 2     SL 7   0,0,z        Immm (2a-d)
```

Space group No. 35 Cmm2

Superlattices

```
SL  0  C(1,1,1)         density  1  add.gen.
SL  1  P(1/2,1/2,1)     density  2  add.gen.  1/2,0,0
SL  4  P(1/4,1/2,1)     density  4  add.gen.  1/4,0,0
SL  5  P(1/2,1/4,1)     density  4  add.gen.  0,1/4,0
```

Wyckoff letter   non-characteristic crystallographic orbits

```
8 f 1      SL 0   x,y,z        Cmmm (8p-q)
           SL 1   x,1/4,z      Pmmm (2i-t)    1/4,y,z     Pmmm (2i-t)
           SL 4   1/8,1/4,z    Pmmm (1a-h)
           SL 5   1/4,1/8,z    Pmmm (1a-h)
4 e m      SL 0   0,y,z        Cmmm (4g-j)
           SL 1   0,1/4,z      Pmmm (1a-h)
4 d m      SL 0   x,0,z        Cmmm (4g-j)
           SL 1   1/4,0,z      Pmmm (1a-h)
4 c 2      SL 1   1/4,1/4,z    Pmmm (1a-h)
2 b mm2    SL 0   0,1/2,z      Cmmm (2a-d)
2 a mm2    SL 0   0,0,z        Cmmm (2a-d)
```

Space group No. 36 Cmc2$_1$

Superlattices

```
SL  0  C(1,1,1)          density  1  add.gen.
SL  1  P(1/2,1/2,1),     density  2  add.gen.  1/2,0,0
SL  2  C(1,1,1/2)        density  2  add.gen.  0,0,1/2
SL  3  F(1,1,1)          density  2  add.gen.  0,1/2,1/2
SL  6  P(1/2,1/2,1/2)    density  4  add.gen.  1/2,0,0  0,0,1/2
SL  7  A(1/2,1/2,1)      density  4  add.gen.  0,1/4,1/2
```

Wyckoff letter   non-characteristic crystallographic orbits

```
8 b 1      SL 0   x,y,z        Cmcm (8g)
           SL 1   1/4,y,z      Pmcm (2e,f)
           SL 2   x,0,z        Cmmm (4g-j)
           SL 3   x,1/4,z      Fmmm (8g-i)
           SL 6   1/4,1/4,z    Pmmm (1a-h)    1/4,0,z     Pmmm (1a-h)
           SL 7   1/4,1/8,z    Ammm (2a-d)
4 a m      SL 0   0,y,z        Cmcm (4c)
           SL 2   0,1/2,z      Cmmm (2a-d)    0,0,z       Cmmm (2a-d)
           SL 3   0,1/4,z      Fmmm (4a,b)
```

Space group No. 37 Ccc2

Superlattices

```
SL  0  C(1,1,1)          density  1  add.gen.
SL  2  C(1,1,1/2)        density  2  add.gen.  0,0,1/2
SL  3  F(1,1,1)          density  2  add.gen.  0,1/2,1/2
SL  6  P(1/2,1/2,1/2)    density  4  add.gen.  1/2,0,0  0,0,1/2
```

Wyckoff letter   non-characteristic crystallographic orbits

```
8 d 1    SL 0  x,y,z       Cccm (8l)
         SL 2  x,0,z       Cmmm (4g-j)    0,y,z      Cmmm (4g-j)
         SL 3  1/4,y,z     Fmmm (8g-i)    x,1/4,z    Fmmm (8g-i)
         SL 6  0,1/4,z     Pmmm (1a-h)    1/4,0,z    Pmmm (1a-h)
4 c 2    SL 3  1/4,1/4,z   Fmmm (4a,b)
4 b 2    SL 2  0,1/2,z     Cmmm (2a-d)
4 a 2    SL 2  0,0,z       Cmmm (2a-d)
```

Space group No. 38 Amm2

Superlattices

```
SL  0  A(1,1,1)          density  1  add.gen.
SL  1  P(1,1/2,1/2)      density  2  add.gen.  0,1/2,0
SL  2  A(1/2,1,1)        density  2  add.gen.  1/2,0,0
SL  4  P(1/2,1/2,1/2)    density  4  add.gen.  1/2,0,0  0,1/2,0
```

Wyckoff letter   non-characteristic crystallographic orbits

```
8 f 1      SL 0  x,y,z       Ammm (8n,o)
           SL 1  x,1/4,z     Pmmm (2i-t)
           SL 2  1/4,y,z     Ammm (4g-j)
           SL 4  1/4,1/4,z   Pmmm (1a-h)
4 e m      SL 0  1/2,y,z     Ammm (4g-j)
           SL 1  1/2,1/4,z   Pmmm (1a-h)
4 d m      SL 0  0,y,z       Ammm (4g-j)
           SL 1  0,1/4,z     Pmmm (1a-h)
4 c m      SL 0  x,0,z       Ammm (4k,l)
           SL 2  1/4,0,z     Ammm (2a-d)
2 b mm2    SL 0  1/2,0,z     Ammm (2a-d)
2 a mm2    SL 0  0,0,z       Ammm (2a-d)
```

Space group   No. 39   Abm2

Superlattices

SL	0	P(1,1,1)	density	1	add.gen.		
SL	1	P(1,1/2,1/2)	density	2	add.gen.	0,1/2,0	
SL	3	F(1,1,1)	density	2	add.gen.	1/2,0,1/2	
SL	4	P(1/2,1/2,1/2)	density	4	add.gen.	1/2,0,0	0,1/2,0
SL	5	P(1,1/4,1/2)	density	4	add.gen.	0,1/4,0	

Wyckoff letter   non-characteristic crystallographic orbits

8 d 1	SL 0	x,y,z	Abmm (8m,n)			
	SL 1	x,0,z	Pmmm (2i-t)	1/2,y,z	Pmmm (2i-t)	
		0,y,z	Pmmm (2i-t)			
	SL 3	1/4,y,z	Fmmm (8g-i)			
	SL 4	1/4,0,z	Pmmm (1a-h)			
	SL 5	1/2,1/8,z	Pmmm (1a-h)	0,1/8,z	Pmmm (1a-h)	
4 c m	SL 0	x,1/4,z	Abmm (4g)			
	SL 1	1/2,1/4,z	Pmmm (1a-h)	0,1/4,z	Pmmm (1a-h)	
	SL 3	1/4,1/4,z	Fmmm (4a,b)			
4 b 2	SL 1	1/2,0,z	Pmmm (1a-h)			
4 a 2	SL 1	0,0,z	Pmmm (1a-h)			

Space group   No. 40   Ama2

Superlattices

SL	0	A(1,1,1)	density	1	add.gen.		
SL	2	A(1/2,1,1)	density	2	add.gen.	1/2,0,0	
SL	3	F(1,1,1)	density	2	add.gen.	1/2,0,1/2	
SL	4	P(1/2,1/2,1/2)	density	4	add.gen.	1/2,0,0	0,1/2,0
SL	11	A(1/4,1,1)	density	4	add.gen.	1/4,0,0	

Wyckoff letter   non-characteristic crystallographic orbits

8 c 1	SL 0	x,y,z	Amam (8f)		
	SL 2	x,0,z	Ammm (4k,l)	0,y,z	Ammm (4g-j)
	SL 3	x,1/4,z	Fmmm (8g-i)		
	SL 4	0,1/4,z	Pmmm (1a-h)		
	SL11	1/8,0,z	Ammm (2a-d)		
4 b m	SL 0	1/4,y,z	Amam (4c)		
	SL 2	1/4,0,z	Ammm (2a-d)		
	SL 3	1/4,1/4,z	Fmmm (4a,b)		
4 a 2	SL 2	0,0,z	Ammm (2a-d)		

Space group   No. 41   Aba2

Superlattices

```
SL   0   A(1,1,1)           density  1   add.gen.
SL   1   P(1,1/2,1/2)       density  2   add.gen.   0,1/2,0
SL   2   A(1/2,1,1)         density  2   add.gen.   1/2,0,0
SL   3   F(1,1,1)           density  2   add.gen.   1/2,0,1/2
Sl   4   P(1/2,1/2,1/2)     density  4   add.gen.   1/2,0,0   0,1/2,0
SL   9   C(1,1/2,1/2)       density  4   add.gen.   1/2,1/4,0
SL  12   F(1/2,1,1)         density  4   add.gen.   1/4,0,1/2
```

Wyckoff letter   non-characteristic crystallographic orbits

```
8 b 1       SL  0   x,y,z        Acam (8f)
            SL  1   1/4,y,z      Pmam (2e,f)
            SL  2   x,1/4,z      Abmm (4g)
            SL  3   0,y,z        Fmmm (8g-i)      x,0,z         Fmmm (8g-i)
            SL  4   1/4,0,z      Pmmm (1a-h)      0,1/4,z       Pmmm (1a-h)
                    1/4,1/4,z    Pmmm (1a-h)
            SL  9   1/4,3/8,z    Cmmm (2a-d)      1/4,1/8,z     Cmmm (2a-d)
            SL 12   1/8,1/4,z    Fmmm (4a,b)
  4 a 2     SL  3   0,0,z        Fmmm (4a,b)
```


Space group   No. 42   Fmm2

Superlattices

```
SL   0   F(1,1,1)           density  1   add.gen.
SL   1   P(1/2,1/2,1/2)     density  2   add.gen.   1/2,0,0
SL   2   P(1/4,1/2,1/2)     density  4   add.gen.   1/4,0,0
SL   3   P(1/2,1/4,1/2)     density  4   add.gen.   0,1/4,0
```

Wyckoff letter   non-characteristic crystallographic orbits

```
16 e 1      SL  0   x,y,z        Fmmm (16m-o)
            SL  1   x,1/4,z      Pmmm (2i-t)      1/4,y,z       Pmmm (2i-t)
            SL  2   1/8,1/4,z    Pmmm (1a-h)
            SL  3   1/4,1/8,z    Pmmm (1a-h)
 8 d m      SL  0   x,0,z        Fmmm (8g-i)
            SL  1   1/4,0,z      Pmmm (1a-h)
 8 c m      SL  0   0,y,z        Fmmm (8g-i)
            SL  1   0,1/4,z      Pmmm (1a-h)
 8 b 2      SL  1   1/4,1/4,z    Pmmm (1a-h)
 4 a mm2    SL  0   0,0,z        Fmmm (4a,b)
```

Space group   No. 43   Fdd2

Superlattices

```
SL  0  F(1,1,1)        density  1  add.gen.
SL  8  I(1/2,1/2,1/2)  density  4  add.gen.  1/4,1/4,1/4
```

Wyckoff letter   non-characteristic crystallographic orbits

```
    16 b 1      SL 0   x,0,z         Fddd (16e-g)    0,y,z        Fddd (16e-g)
                       1/8,1/8,1/8   Fddd (16c,d)    5/8,5/8,5/8  Fddd (16c,d)
                SL 8   1/4,0,z       Immm (2a-d)
     8 a 2      SL 0   0,0,z         Fddd (8a,b)
```


Space group   No. 44   Imm2

Superlattices

```
SL  0  I(1,1,1)        density  1  add.gen.
SL  1  A(1/2,1,1)      density  2  add.gen.  1/2,0,0
SL  2  B(1,1/2,1)      density  2  add.gen.  0,1/2,0
SL  4  P(1/2,1/2,1/2)  density  4  add.gen.  1/2,0,0  0,1/2,0
```

Wyckoff letter   non-characteristic crystallographic orbits

```
     8 e 1      SL 0   x,y,z         Immm (8l-n)
                SL 1   1/4,y,z       Ammm (4g-j)
                SL 2   x,1/4,z       Bmmm (4g-j)
                SL 4   1/4,1/4,z     Pmmm (1a-h)
     4 d m      SL 0   0,y,z         Immm (4e-j)
                SL 2   0,1/4,z       Bmmm (2a-d)
     4 c m      SL 0   x,0,z         Immm (4e-j)
                SL 1   1/4,0,z       Ammm (2a-d)
     2 b mm2    SL 0   0,1/2,z       Immm (2a-d)
     2 a mm2    SL 0   0,0,z         Immm (2a-d)
```

Space group   No. 45   Iba2

Superlattices

```
SL  0   I(1,1,1)          density  1   add.gen.
SL  1   A(1/2,1,1)        density  2   add.gen.   1/2,0,0
SL  2   B(1,1/2,1)        density  2   add.gen.   0,1/2,0
SL  3   C(1,1,1/2)        density  2   add.gen.   0,0,1/2
SL  4   P(1/2,1/2,1/2)    density  4   add.gen.   1/2,0,0   0,1/2,0
SL  6   F(1/2,1,1)        density  4   add.gen.   1/4,0,1/2
SL  8   F(1,1/2,1)        density  4   add.gen.   0,1/4,1/2
```

Wyckoff letter   non-characteristic crystallographic orbits

```
 8 c 1      SL 0   x,y,z        Ibam (8j)
            SL 1   x,1/4,z      Abmm (4g)
            SL 2   1/4,y,z      Bmam (4g)
            SL 3   0,y,z        Cmmm (4g-j)    x,0,z         Cmmm (4g-j)
            SL 4   1/4,1/4,z    Pmmm (1a-h)    0,1/4,z       Pmmm (1a-h)
                   1/4,0,z      Pmmm (1a-h)
            SL 6   1/8,1/4,z    Fmmm (4a,b)
            SL 8   1/4,1/8,z    Fmmm (4a,b)
 4 b 2      SL 3   0,1/2,z      Cmmm (2a-d)
 4 a 2      SL 3   0,0,z        Cmmm (2a-d)
```

Space group   No. 46   Ima2

Superlattices

```
SL  0   I(1,1,1)          density  1   add.gen.
SL  1   A(1/2,1,1)        density  2   add.gen.   1/2,0,0
SL  3   C(1,1,1/2)        density  2   add.gen.   0,0,1/2
SL  4   P(1/2,1/2,1/2)    density  4   add.gen.   1/2,0,0   0,1/2,0
SL  5   A(1/4,1,1)        density  4   add.gen.   1/4,0,0
```

Wyckoff letter   non-characteristic crystallographic orbits

```
 8 c 1      SL 0   x,y,z        Imam (8h,i)
            SL 1   x,0,z        Ammm (4k,l)    0,y,z         Ammm (4g-j)
            SL 3   x,1/4,z      Cmmm (4g-j)
            SL 4   0,1/4,z      Pmmm (1a-h)
            SL 5   1/8,0,z      Ammm (2a-d)
 4 b m      SL 0   1/4,y,z      Imcm (4e)
            SL 1   1/4,0,z      Ammm (2a-d)
            SL 3   1/4,1/4,z    Cmmm (2a-d)    1/4,3/4,z     Cmmm (2a-d)
 4 a 2      SL 1   0,0,z        Ammm (2a-d)
```

Space group   No. 47   Pmmm

Superlattices

SL	1	P(1/2,1,1)	density	2	add.gen.	1/2,0,0		
SL	2	P(1,1/2,1)	density	2	add.gen.	0,1/2,0		
SL	3	P(1,1,1/2)	density	2	add.gen.	0,0,1/2		
SL	9	P(1/2,1/2,1)	density	4	add.gen.	1/2,0,0	0,1/2,0	
SL	10	P(1/2,1,1/2)	density	4	add.gen.	1/2,0,0	0,0,1/2	
SL	16	P(1,1/2,1/2)	density	4	add.gen.	0,1/2,0	0,0,1/2	
SL	35	P(1/2,1/2,1/2)	density	8	add.gen.	1/2,0,0	0,1/2,0	0,0,1/2

Wyckoff letter   non-characteristic crystallographic orbits

8 α 1	SL 1	1/4,y,z	Pmmm	(4u-z)
	SL 2	x,1/4,z	Pmmm	(4u-z)
	SL 3	x,y,1/4	Pmmm	(4u-z)
	SL 9	1/4,1/4,z	Pmmm	(2i-t)
	SL10	1/4,y,1/4	Pmmm	(2i-t)
	SL16	x,1/4,1/4	Pmmm	(2i-t)
	SL35	1/4,1/4,1/4	Pmmm	(1a-h)
4 z m	SL 1	1/4,y,1/2	Pmmm	(2i-t)
	SL 2	x,1/4,1/2	Pmmm	(2i-t)
	SL 9	1/4,1/4,1/2	Pmmm	(1a-h)
4 y m	SL 1	1/4,y,0	Pmmm	(2i-t)
	SL 2	x,1/4,0	Pmmm	(2i-t)
	SL 9	1/4,1/4,0	Pmmm	(1a-h)
4 x m	SL 1	1/4,1/2,z	Pmmm	(2i-t)
	SL 3	x,1/2,1/4	Pmmm	(2i-t)
	SL10	1/4,1/2,1/4	Pmmm	(1a-h)
4 w m	SL 1	1/4,0,z	Pmmm	(2i-t)
	SL 3	x,0,1/4	Pmmm	(2i-t)
	SL10	1/4,0,1/4	Pmmm	(1a-h)
4 v m	SL 2	1/2,1/4,z	Pmmm	(2i-t)
	SL 3	1/2,y,1/4	Pmmm	(2i-t)
	SL16	1/2,1/4,1/4	Pmmm	(1a-h)
4 u m	SL 2	0,1/4,z	Pmmm	(2i-t)
	SL 3	0,y,1/4	Pmmm	(2i-t)
	SL16	0,1/4,1/4	Pmmm	(1a-h)
2 t mm2	SL 3	1/2,1/2,1/4	Pmmm	(1a-h)
2 s mm2	SL 3	1/2,0,1/4	Pmmm	(1a-h)
2 r mm2	SL 3	0,1/2,1/4	Pmmm	(1a-h)
2 q mm2	SL 3	0,0,1/4	Pmmm	(1a-h)
2 p mm2	SL 2	1/2,1/4,1/2	Pmmm	(1a-h)
2 o mm2	SL 2	1/2,1/4,0	Pmmm	(1a-h)
2 n mm2	SL 2	0,1/4,1/2	Pmmm	(1a-h)
2 m mm2	SL 2	0,1/4,0	Pmmm	(1a-h)
2 l mm2	SL 1	1/4,1/2,1/2	Pmmm	(1a-h)
2 k mm2	SL 1	1/4,1/2,0	Pmmm	(1a-h)
2 j mm2	SL 1	1/4,0,1/2	Pmmm	(1a-h)
2 i mm2	SL 1	1/4,0,0	Pmmm	(1a-h)

Space group No. 48 Pnnn

Superlattices

```
SL   4   A(1,1,1)         density  2  add.gen.  0,1/2,1/2
SL   5   B(1,1,1)         density  2  add.gen.  1/2,0,1/2
SL   6   C(1,1,1)         density  2  add.gen.  1/2,1/2,0
SL   7   I(1,1,1)         density  2  add.gen.  1/2,1/2,1/2
SL  11   A(1/2,1,1)       density  4  add.gen.  1/2,0,0   0,1/2,1/2
SL  18   B(1,1/2,1)       density  4  add.gen.  0,1/2,0   1/2,0,1/2
SL  24   C(1,1,1/2)       density  4  add.gen.  0,0,1/2   1/2,1/2,0
SL  26   F(1,1,1)         density  4  add.gen.  0,1/2,1/2  1/2,0,1/2
SL  35   P(1/2,1/2,1/2)   density  8  add.gen.  1/2,0,0   0,1/2,0   0,0,1/2
```

Wyckoff letter   non-characteristic crystallographic orbits

```
8 m 1     SL 4   0,y,z         Amaa (8l)
          SL 5   x,0,z         Bbmb (8l)
          SL 6   x,y,0         Cccm (8l)
          SL 7   x,y,1/4       Immm (8l-n)    x,1/4,z      Immm (8l-n)
                 1/4,y,z       Immm (8l-n)
          SL11   0,1/4,z       Ammm (4g-j)    0,y,1/4      Ammm (4g-j)
          SL18   1/4,0,z       Bmmm (4g-j)    x,0,1/4      Bmmm (4g-j)
          SL24   1/4,y,0       Cmmm (4g-j)    x,1/4,0      Cmmm (4g-j)
          SL26   0,0,z         Fmmm (8g-i)    0,y,0        Fmmm (8g-i)
                 x,0,0         Fmmm (8g-i)
          SL35   1/4,0,0       Pmmm (1a-h)    0,0,1/4      Pmmm (1a-h)
                 0,1/4,0       Pmmm (1a-h)
4 l 2     SL 7   1/4,3/4,z     Immm (4e-j)
          SL24   1/4,3/4,0     Cmmm (2a-d)
4 k 2     SL 7   1/4,1/4,z     Immm (4e-j)
          SL24   1/4,1/4,0     Cmmm (2a-d)
4 j 2     SL 7   3/4,y,1/4     Immm (4e-j)
          SL18   3/4,0,1/4     Bmmm (2a-d)
4 i 2     SL 7   1/4,y,1/4     Immm (4e-j)
          SL18   1/4,0,1/4     Bmmm (2a-d)
4 h 2     SL 7   x,1/4,3/4     Immm (·4e-j)
          SL11   0,1/4,3/4     Ammm (2a-d)
4 g 2     SL 7   x,1/4,1/4     Immm (4e-j)
          SL11   0,1/4,1/4     Ammm (2a-d)
4 f 1̄     SL26   0,0,0         Fmmm (4a,b)
4 e 1̄     SL26   1/2,1/2,1/2   Fmmm (4a,b)
2 d 222   SL 7   1/4,3/4,1/4   Immm (2a-d)
2 c 222   SL 7   1/4,1/4,3/4   Immm (2a-d)
2 b 222   SL 7   3/4,1/4,1/4   Immm (2a-d)
2 a 222   SL 7   1/4,1/4,1/4   Immm (2a-d)
```

Space group No. 49 Pccm

Superlattices

```
SL  3   P(1,1,1/2)      density  2   add.gen.  0,0,1/2
SL  4   A(1,1,1)        density  2   add.gen.  0,1/2,1/2
SL  5   B(1,1,1)        density  2   add.gen.  1/2,0,1/2
Sl 10   P(1/2,1,1/2)    density  4   add.gen.  1/2,0,0   0,0,1/2
SL 16   P(1,1/2,1/2)    density  4   add.gen.  0,1/2,0   0,0,1/2
SL 21   P(1,1,1/4)      density  4   add.gen.  0,0,1/4
SL 26   F(1,1,1)        density  4   add.gen.  0,1/2,1/2  1/2,0,1/2
SL 35   P(1/2,1/2,1/2)  density  8   add.gen.  1/2,0,0   0,1/2,0   0,0,1/2
SL 40   P(1/2,1,1/4)    density  8   add.gen.  1/2,0,0   0,0,1/4
SL 52   P(1,1/2,1/4)    density  8   add.gen.  0,1/2,0   0,0,1/4
```

Wyckoff letter   non-characteristic crystallographic orbits

```
8 r 1    SL 3   x,y,1/4      Pmmm (4u-z)    1/2,y,z     Pmmm (4u-z)
                0,y,z        Pmmm (4u-z)    x,1/2,z     Pmmm (4u-z)
                x,0,z        Pmmm (4u-z)
         SL 4   x,1/4,z      Abmm (8m,n)
         SL 5   1/4,y,z      Bmam (8m,n)
         SL10   1/4,y,1/4    Pmmm (2i-t)    1/4,1/2,z   Pmmm (2i-t)
                1/4,0,z      Pmmm (2i-t)
         SL16   x,1/4,1/4    Pmmm (2i-t)    1/2,1/4,z   Pmmm (2i-t)
                0,1/4,z      Pmmm (2i-t)
         SL21   x,1/2,1/8    Pmmm (2i-t)    x,0,1/8     Pmmm (2i-t)
                1/2,y,1/8    Pmmm (2i-t)    0,y,1/8     Pmmm (2i-t)
         SL26   1/4,1/4,z    Fmmm (8g-i)
         SL35   1/4,1/4,1/4  Pmmm (1a-h)
         SL40   1/4,1/2,1/8  Pmmm (1a-h)    1/4,0,1/8   Pmmm (1a-h)
         SL52   1/2,1/4,1/8  Pmmm (1a-h)    0,1/4,1/8   Pmmm (1a-h)
4 q .m   SL 3   1/2,y,0      Pmmm (2i-t)    0,y,0       Pmmm (2i-t)
                x,1/2,0      Pmmm (2i-t)    x,0,0       Pmmm (2i-t)
         SL 4   x,1/4,0      Abmm (4g)
         SL 5   1/4,y,0      Bmam (4g)
         SL10   1/4,1/2,0    Pmmm (1a-h)    1/4,0,0     Pmmm (1a-h)
         SL16   1/2,1/4,0    Pmmm (1a-h)    0,1/4,0     Pmmm (1a-h)
         SL26   1/4,3/4,0    Fmmm (4a,b)    1/4,1/4,0   Fmmm (4a,b)
4 p 2    SL 3   1/2,0,z      Pmmm (2i-t)
         SL21   1/2,0,1/8    Pmmm (1a-h)
4 o 2    SL 3   0,1/2,z      Pmmm (2i-t)
         SL21   0,1/2,1/8    Pmmm (1a-h)
4 n 2    SL 3   1/2,1/2,z    Pmmm (2i-t)
         SL21   1/2,1/2,1/8  Pmmm (1a-h)
4 m 2    SL 3   0,0,z        Pmmm (2i-t)
         SL21   0,0,1/8      Pmmm (1a-h)
4 l 2    SL 3   1/2,y,1/4    Pmmm (2i-t)
         SL16   1/2,1/4,1/4  Pmmm (1a-h)
4 k 2    SL 3   0,y,1/4      Pmmm (2i-t)
         SL16   0,1/4,1/4    Pmmm (1a-h)
4 j 2    SL 3   x,1/2,1/4    Pmmm (2i-t)
         SL10   1/4,1/2,1/4  Pmmm (1a-h)
4 i 2    SL 3   x,0,1/4      Pmmm (2i-t)
         SL10   1/4,0,1/4    Pmmm (1a-h)
```

(continued on next page)

Space group No. 49 Pccm   (continued)

Wyckoff letter   non-characteristic crystallographic orbits

```
2 h 222    SL 3   1/2,1/2,1/4   Pmmm (1a-h)
2 g 222    SL 3   0,1/2,1/4     Pmmm (1a-h)
2 f 222    SL 3   1/2,0,1/4     Pmmm (1a-h)
2 e 222    SL 3   0,0,1/4       Pmmm (1a-h)
2 d 2/m    SL 3   1/2,0,0       Pmmm (1a-h)
2 c 2/m    SL 3   0,1/2,0       Pmmm (1a-h)
2 b 2/m    SL 3   1/2,1/2,0     Pmmm (1a-h)
2 a 2/m    SL 3   0,0,0         Pmmm (1a-h)
```

Space group No. 50 Pban

Superlattices

```
SL  1  P(1/2,1,1)      density  2  add.gen.  1/2,0,0
SL  2  P(1,1/2,1)      density  2  add.gen.  0,1/2,0
SL  6  C(1,1,1)        density  2  add.gen.  1/2,1/2,0
SL  7  I(1,1,1)        density  2  add.gen.  1/2,1/2,1/2
SL  9  P(1/2,1/2,1)    density  4  add.gen.  1/2,0,0   0,1/2,0
SL 11  A(1/2,1,1)      density  4  add.gen.  1/2,0,0   0,1/2,1/2
SL 13  C(1/2,1,1)      density  4  add.gen.  1/4,1/2,0
SL 18  B(1,1/2,1)      density  4  add.gen.  0,1/2,0   1/2,0,1/2
SL 19  C(1,1/2,1)      density  4  add.gen.  1/2,1/4,0
SL 24  C(1,1,1/2)      density  4  add.gen.  0,0,1/2   1/2,1/2,0
SL 28  P(1/4,1/2,1)    density  8  add.gen.  1/4,0,0   0,1/2,0
SL 34  P(1/2,1/4,1)    density  8  add.gen.  1/2,0,0   0,1/4,0
SL 35  P(1/2,1/2,1/2)  density  8  add.gen.  1/2,0,0   0,1/2,0   0,0,1/2
SL 45  F(1/2,1,1)      density  8  add.gen.  0,1/2,1/2  1/4,0,1/2
SL 57  F(1,1/2,1)      density  8  add.gen.  0,1/4,1/2  1/2,0,1/2
```

Wyckoff letter   non-characteristic crystallographic orbits

```
  8 m 1   SL  1  x,0,z        Pbmb (4q)
          SL  2  0,y,z        Pmaa (4q)
          SL  6  x,1/4,z      Cmmm (8n,o)      1/4,y,z      Cmmm (8n,o)
                 x,y,1/2      Cmmm (8p,q)      x,y,0        Cmmm (8p-q)
          SL  7  x,y,1/4      Ibam (8j)
          SL  9  0,0,z        Pmmm (2i-t)      x,0,1/2      Pmmm (2i-t)
                 x,0,0        Pmmm (2i-t)      1/4,0,z      Pmmm (2i-t)
                 0,y,1/2      Pmmm (2i-t)      0,y,0        Pmmm (2i-t)
                 0,1/4,z      Pmmm (2i-t)
          SL11   x,0,1/4      Abmm (4g)
          SL13   1/8,0,z      Cmma (4g)
          SL18   0,y,1/4      Bmam (4g)
          SL19   0,1/8,z      Cmma (4g)
          SL24   1/4,y,1/4    Cmmm (4g-j)      x,1/4,1/4    Cmmm (4g-j)
          SL28   1/8,0,1/2    Pmmm (1a-h)      1/8,0,0      Pmmm (1a-h)
          SL34   0,1/8,1/2    Pmmm (1a-h)      0,1/8,0      Pmmm (1a-h)
          SL35   0,0,1/4      Pmmm (1a-h)      1/4,0,1/4    Pmmm (1a-h)
                 0,1/4,1/4    Pmmm (1a-h)
          SL45   1/8,1/2,1/4  Fmmm (4a,b)      1/8,0,1/4    Fmmm (4a,b)
          SL57   1/2,1/8,1/4  Fmmm (4a,b)      0,1/8,1/4    Fmmm (4a,b)
  4 l 2   SL  6  1/4,3/4,z    Cmmm (4k,l)
          SL24   1/4,3/4,1/4  Cmmm (2a-d)
  4 k 2   SL  6  1/4,1/4,z    Cmmm (4k,l)
          SL24   1/4,1/4,1/4  Cmmm (2a-d)
  4 j 2   SL  6  1/4,y,1/2    Cmmm (4g-j)
          SL  9  1/4,0,1/2    Pmmm (1a-h)
  4 i 2   SL  6  1/4,y,0      Cmmm (4g-j)
          SL  9  1/4,0,0      Pmmm (1a-h)
  4 h 2   SL  6  x,1/4,1/2    Cmmm (4g-j)
          SL  9  0,1/4,1/2    Pmmm (1a-h)
```

(continued on next page)

space group No. 50 Pban   (continued)

Wyckoff letter   non-characteristic crystallographic orbits

```
4 g 2      SL 6   x,1/4,0      Cmmm (4g-j)
           SL 9   0,1/4,0      Pmmm (1a-h)
4 f 1̄      SL 9   0,0,1/2      Pmmm (1a-h)
4 e 1̄      SL 9   0,0,0        Pmmm (1a-h)
2 d 222    SL 6   1/4,1/4,1/2  Cmmm (2a-d)
2 c 222    SL 6   3/4,1/4,1/2  Cmmm (2a-d)
2 b 222    SL 6   3/4,1/4,0    Cmmm (2a-d)
2 a 222    SL 6   1/4,1/4,0    Cmmm (2a-d)
```

Space group No. 51 Pmma

Superlattices

```
SL  1  P(1/2,1,1)     density  2  add.gen.  1/2,0,0
SL  2  P(1,1/2,1)     density  2  add.gen.  0,1/2,0
SL  5  B(1,1,1)       density  2  add.gen.  1/2,0,1/2
SL  8  P(1/4,1,1)     density  4  add.gen.  1/4,0,0
SL  9  P(1/2,1/2,1)   density  4  add.gen.  1/2,0,0  0,1/2,0
SL 10  P(1/2,1,1/2)   density  4  add.gen.  1/2,0,0  0,0,1/2
SL 18  B(1,1/2,1)     density  4  add.gen.  0,1/2,0  1/2,0,1/2
SL 28  P(1/4,1/2,1)   density  8  add.gen.  1/4,0,0  0,1/2,0
SL 35  P(1/2,1/2,1/2) density  8  add.gen.  1/2,0,0  0,1/2,0  0,0,1/2
```

Wyckoff letter   non-characteristic crystallographic orbits

Wyckoff	SL	orbit	group	orbit	group
8 l 1	SL 1	0,y,z	Pmmm (4u-z)	x,y,1/2	Pmmm (4u-z)
		x,y,0	Pmmm (4u-z)		
	SL 2	x,1/4,z	Pmma (4i,j)		
	SL 5	x,y,1/4	Bmmm (8n,o)		
	SL 8	1/8,y,1/2	Pmmm (2i-t)	1/8,y,0	Pmmm (2i-t)
	SL 9	0,1/4,z	Pmmm (2i-t)	x,1/4,1/2	Pmmm (2i-t)
		x,1/4,0	Pmmm (2i-t)		
	SL10	0,y,1/4	Pmmm (2i-t)		
	SL18	x,1/4,1/4	Bmmm (4g-j)		
	SL28	1/8,1/4,1/2	Pmmm (1a-h)	1/8,1/4,0	Pmmm (1a-h)
	SL35	0,1/4,1/4	Pmmm (1a-h)		
4 k m	SL 1	1/4,y,1/2	Pmmm (2i-t)	1/4,y,0	Pmmm (2i-t)
	SL 2	1/4,1/4,z	Pmma (2e,f)		
	SL 5	1/4,y,1/4	Bmmm (4k,l)	1/4,y,3/4	Bmmm (4k,l)
	SL 9	1/4,1/4,1/2	Pmmm (1a-h)	1/4,1/4,0	Pmmm (1a-h)
	SL18	1/4,1/4,1/4	Bmmm (2a-d)	1/4,1/4,3/4	Bmmm (2a-d)
4 j m	SL 1	0,1/2,z	Pmmm (2i-t)	x,1/2,1/2	Pmmm (2i-t)
		x,1/2,0	Pmmm (2i-t)		
	SL 5	x,1/2,1/4	Bmmm (4g-j)		
	SL 8	1/8,1/2,1/2	Pmmm (1a-h)	1/8,1/2,0	Pmmm (1a-h)
	SL10	0,1/2,1/4	Pmmm (1a-h)		
4 ı m	SL 1	0,0,z	Pmmm (2i-t)	x,0,1/2	Pmmm (2i-t)
		x,0,0	Pmmm (2i-t)		
	SL 5	x,0,1/4	Bmmm (4g-j)		
	SL 8	1/8,0,1/2	Pmmm (1a-h)	1/8,0,0	Pmmm (1a-h)
	SL10	0,0,1/4	Pmmm (1a-h)		
4 h 2	SL 1	0,y,1/2	Pmmm (2i-t)		
	SL 9	0,1/4,1/2	Pmmm (1a-h)		
4 g 2	SL 1	0,y,0	Pmmm (2i-t)		
	SL 9	0,1/4,0	Pmmm (1a-h)		
2 f mm2	SL 1	1/4,1/2,1/2	Pmmm (1a-h)	1/4,1/2,0	Pmmm (1a-h)
	SL 5	1/4,1/2,1/4	Bmmm (2a-d)	1/4,1/2,3/4	Bmmm (2a-d)
2 e mm2	SL 1	1/4,0,1/2	Pmmm (1a-h)	1/4,0,0	Pmmm (1a-h)
	SL 5	1/4,0,1/4	Bmmm (2a-d)	1/4,0,3/4	Bmmm (2a-d)
2 d 2/m	SL 1	0,1/2,1/2	Pmmm (1a-h)		
2 c 2/m	SL 1	0,0,1/2	Pmmm (1a-h)		
2 b 2/m	SL 1	0,1/2,0	Pmmm (1a-h)		
2 a 2/m	SL 1	0,0,0	Pmmm (1a-h)		

Space group No. 52 Pnna

Superlattices

```
SL   1   P(1/2,1,1)       density  2  add.gen.  1/2,0,0
SL   4   A(1,1,1)         density  2  add.gen.  0,1/2,1/2
Sl   5   B(1,1,1)         density  2  add.gen.  1/2,0,1/2
SL   7   I(1,1,1)         density  2  add.gen.  1/2,1/2,1/2
SL  10   P(1/2,1,1/2)     density  4  add.gen.  1/2,0,0   0,0,1/2
SL  11   A(1/2,1,1)       density  4  add.gen.  1/2,0,0   0,1/2,1/2
SL  14   I(1/2,1,1)       density  4  add.gen.  1/4,1/2,1/2
SL  18   B(1,1/2,1)       density  4  add.gen.  0,1/2,0   1/2,0,1/2
SL  26   F(1,1,1)         density  4  add.gen.  0,1/2,1/2  1/2,0,1/2
SL  30   A(1/4,1,1)       density  8  add.gen.  1/4,0,0   0,1/2,1/2
SL  35   P(1/2,1/2,1/2)   density  8  add.gen.  1/2,0,0   0,1/2,0   0,0,1/2
SL  43   C(1/2,1,1/2)     density  8  add.gen.  0,0,1/2   1/4,1/2,0
SL  49   B(1,1/4,1)       density  8  add.gen.  0,1/4,0   1/2,0,1/2
```

Wyckoff letter   non-characteristic crystallographic orbits

```
8 e 1      SL 1   x,y,0           Pncm (4h)
           SL 4   0,y,z           Amaa (8l)
           SL 5   x,y,1/4         Bbmm (8f)       x,1/4,z         Bbmm (8g)
           SL 7   1/4,y,z         Imma (8h,i)     x,0,z          Imma (8h,i)
           SL10   x,1/4,0         Pbmm (2e,f)
           SL11   0,y,0           Ammm (4g-j)     x,0,1/2        Ammm (4k,l)
                  x,0,0           Ammm (4k,l)     1/4,y,0        Ammm (4g-j)
                  0,0,z           Ammm (4g-j)
           SL14   1/8,y,0         Imam (4e)
           SL18   3/4,y,1/4       Bmmm (4k,l)     1/4,y,1/4      Bmmm (4k,l)
                  1/4,1/4,z       Bmmm (4g-j)     x,0,1/4        Bmmm (4g-j)
           SL26   0,y,1/4         Fmmm (8g-i)     0,1/4,z        Fmmm (8g-i)
           SL30   1/8,0,1/2       Ammm (2a-d)     1/8,0,0        Ammm (2a-d)
           SL35   0,1/4,0         Pmmm (1a-h)     1/4,1/4,0      Pmmm (1a-h)
                  0,0,1/4         Pmmm (1a-h)
           SL43   1/8,3/4,0       Cmmm (2a-d)     1/8,1/4,0      Cmmm (2a-d)
           SL49   3/4,1/8,1/4     Bmmm (2a-d)     1/4,1/8,1/4    Bmmm (2a-d)
4 d 2      SL 5   x,1/4,1/4       Bbmm (4c)
           SL18   3/4,1/4,1/4     Bmmm (2a-d)     1/4,1/4,1/4    Bmmm (2a-d)
           SL26   0,1/4,1/4       Fmmm (4a,b)     1/2,1/4,1/4    Fmmm (4a,b)
4 c 2      SL 7   1/4,0,z         Imma (4e)
           SL11   1/4,0,1/2       Ammm (2a-d)     1/4,0,0        Ammm (2a-d)
           SL18   1/4,0,3/4       Bmmm (2a-d)     1/4,0,1/4      Bmmm (2a-d)
4 b 1̄      SL11   0,0,1/2         Ammm (2a-d)
4 a 1̄      SL11   0,0,0           Ammm (2a-d)
```

Space group   No. 53   Pmna

Superlattices

```
SL  1   P(1/2,1,1)        density  2  add.gen.  1/2,0,0
SL  5   B(1,1,1)          density  2  add.gen.  1/2,0,1/2
SL  7   I(1,1,1)          density  2  add.gen.  1/2,1/2,1/2
SL  8   P(1/4,1,1)        density  4  add.gen.  1/4,0,0
SL 10   P(1/2,1,1/2)      density  4  add.gen.  1/2,0,0    0,0,1/2
SL 11   A(1/2,1,1)        density  4  add.gen.  1/2,0,0    0,1/2,1/2
SL 18   B(1,1/2,1)        density  4  add.gen.  0,1/2,0    1/2,0,1/2
SL 29   P(1/4,1,1/2)      density  8  add.gen.  1/4,0,0    0,0,1/2
SL 30   A(1/4,1,1).       density  8  add.gen.  1/4,0,0    0,1/2,1/2
SL 35   P(1/2,1/2,1/2)    density  8  add.gen.  1/2,0,0    0,1/2,0    0,0,1/2
SL 40   P(1/2,1,1/4)      density  8  add.gen.  1/2,0,0    0,0,1/4
```

Wyckoff letter   non-characteristic crystallographic orbits

```
 8 i 1    SL  1   1/4,y,z        Pmcm (4i,j)      x,y,1/4        Pmcm (4k)
          SL  5   x,1/2,z        Bmmm (8p,q)      x,0,z          Bmmm (8p,q)
                  x,y,0          Bmmm (8n,o)
          SL  7   x,1/4,z        Imma (8h,i)
          SL  8   1/8,y,1/4      Pmcm (2e,f)
          SL10   1/4,y,0         Pmmm (2i-t)      1/4,1/2,z      Pmmm (2i-t)
                 1/4,0,z         Pmmm (2i-t)      x,1/2,1/4      Pmmm (2i-t)
                 x,0,1/4         Pmmm (2i-t)
          SL11   1/4,1/4,z       Ammm (4g-j)      x,3/4,1/4      Ammm (4k,l)
                 x,1/4,1/4       Ammm (4k,l)
          SL18   x,1/4,0         Bmmm (4g-j)
          SL29   1/8,1/2,1/4     Pmmm (1a-h)      1/8,0,1/4      Pmmm (1a-h)
          SL30   1/8,3/4,1/4     Ammm (2a-d)      1/8,1/4,1/4    Ammm (2a-d)
          SL35   1/4,1/4,0       Pmmm (1a-h)
          SL40   1/4,1/2,1/8     Pmmm (1a-h)      1/4,0,1/8      Pmmm (1a-h)
 4 h m    SL  1   0,y,1/4        Pmcm (2e,f)
          SL  5   0,1/2,z        Bmmm (4g-j)      0,0,z          Bmmm (4g-j)
                  0,y,0          Bmmm (4k,l)      0,y,1/2        Bmmm (4k,l)
          SL  7   0,1/4,z        Imma (4e)
          SL10   0,1/2,1/4       Pmmm (1a-h)      0,0,1/4        Pmmm (1a-h)
          SL11   0,3/4,1/4       Ammm (2a-d)      0,1/4,1/4      Ammm (2a-d)
          SL18   0,1/4,0         Bmmm (2a-d)      0,1/4,1/2      Bmmm (2a-d)
 4 g 2    SL  1   1/4,y,1/4      Pmcm (2e,f)
          SL10   1/4,1/2,1/4     Pmmm (1a-h)      1/4,0,1/4      Pmmm (1a-h)
          SL11   1/4,3/4,1/4     Ammm (2a-d)      1/4,1/4,1/4    Ammm (2a-d)
 4 f 2    SL  5   x,1/2,0        Bmmm (4g-j)
          SL10   1/4,1/2,0       Pmmm (1a-h)
 4 e 2    SL  5   x,0,0          Bmmm (4g-j)
          SL10   1/4,0,0         Pmmm (1a-h)
 2 d 2/m  SL  5   0,1/2,0        Bmmm (2a-d)
 2 c 2/m  SL  5   1/2,1/2,0      Bmmm (2a-d)
 2 b 2/m  SL  5   1/2,0,0        Bmmm (2a-d)
 2 a 2/m  SL  5   0,0,0          Bmmm (2a-d)
```

Space group   No. 54   Pcca

Superlattices

```
SL  1   P(1/2,1,1)       density  2   add.gen.  1/2,0,0
SL  3   P(1,1,1/2)       density  2   add.gen.  0,0,1/2
SL  4   A(1,1,1)         density  2   add.gen.  0,1/2,1/2
SL  5   B(1,1,1)         density  2   add.gen.  1/2,0,1/2
SL 10   P(1/2,1,1/2)     density  4   add.gen.  1/2,0,0   0,0,1/2
SL 11   A(1/2,1,1)       density  4   add.gen.  1/2,0,0   0,1/2,1/2
SL 12   B(1/2,1,1)       density  4   add.gen.  1/4,0,1/2
SL 16   P(1,1/2,1/2)     density  4   add.gen.  0,1/2,0   0,0,1/2
SL 23   B(1,1,1/2)       density  4   add.gen.  1/2,0,1/4
SL 26   F(1,1,1)         density  4   add.gen.  0,1/2,1/2   1/2,0,1/2
SL 29   P(1/4,1,1/2)     density  8   add.gen.  1/4,0,0   0,0,1/2
SL 35   P(1/2,1/2,1/2)   density  8   add.gen.  1/2,0,0   0,1/2,0   0,0,1/2
SL 40   P(1/2,1,1/4)     density  8   add.gen.  1/2,0,0   0,0,1/4
SL 45   F(1/2,1,1)       density  8   add.gen.  0,1/2,1/2   1/4,0,1/2
SL 54   B(1,1/2,1/2)     density  8   add.gen.  0,1/2,0   1/2,0,1/4
```

Wyckoff letter   non-characteristic crystallographic orbits

```
  8 f 1      SL  1   x,y,0          Pccm (4q)
             SL  3   1/4,y,z        Pmma (4k)       x,1/2,z         Pmma (4i,j)
                     x,0,z          Pmma (4i,j)
             SL  4   x,1/4,z        Abma (8f)
             SL  5   x,y,1/4        Bmcm (8m,n)     0,y,z           Bmcm (8m,n)
             SL 10   1/4,y,1/4      Pmmm (2i-t)     1/4,y,0         Pmmm (2i-t)
                     x,1/2,0        Pmmm (2i-t)     x,0,0           Pmmm (2i-t)
                     0,y,0          Pmmm (2i-t)     x,1/2,1/4       Pmmm (2i-t)
                     x,0,1/4        Pmmm (2i-t)     0,1/2,z         Pmmm (2i-t)
                     0,0,z          Pmmm (2i-t)
             SL 11   x,1/4,0        Acmm (4g)
             SL 12   1/8,y,0        Bmcm (4g)
             SL 16   1/4,1/4,z      Pmma (2e,f)
             SL 23   1/4,y,3/8      Bmmm (4k,l)     1/4,y,1/8       Bmmm (4k,l)
                     x,1/2,1/8      Bmmm (4g-j)     x,0,1/8         Bmmm (4g-j)
             SL 26   x,1/4,1/4      Fmmm (8g-i)     0,1/4,z         Fmmm (8g-i)
             SL 29   1/8,1/2,0      Pmmm (1a-h)     1/8,0,0         Pmmm (1a-h)
                     1/8,1/2,1/4    Pmmm (1a-h)     1/8,0,1/4       Pmmm (1a-h)
             SL 35   1/4,1/4,1/4    Pmmm (1a-h)     0,1/4,0         Pmmm (1a-h)
                     1/4,1/4,0      Pmmm (1a-h)
             SL 40   0,1/2,1/8      Pmmm (1a-h)     0,0,1/8         Pmmm (1a-h)
             SL 45   1/8,1/4,1/2    Fmmm (4a,b)     1/8,1/4,0       Fmmm (4a,b)
             SL 54   1/4,1/4,3/8    Bmmm (2a-d)     1/4,1/4,1/8     Bmmm (2a-d)
  4 e 2      SL  3   1/4,1/2,z      Pmma (2e,f)
             SL 10   1/4,1/2,0      Pmmm (1a-h)     1/4,1/2,1/4     Pmmm (1a-h)
             SL 23   1/4,1/2,3/8    Bmmm (2a-d)     1/4,1/2,1/8     Bmmm (2a-d)
  4 d 2      SL  3   1/4,0,z        Pmma (2e,f)
             SL 10   1/4,0,0        Pmmm (1a-h)     1/4,0,1/4       Pmmm (1a-h)
             SL 23   1/4,0,3/8      Bmmm (2a-d)     1/4,0,1/8       Bmmm (2a-d)
  4 c 2      SL  5   0,y,1/4        Bmcm (4g)
             SL 10   0,1/2,1/4      Pmmm (1a-h)     0,0,1/4         Pmmm (1a-h)
             SL 26   0,3/4,1/4      Fmmm (4a,b)     0,1/4,1/4       Fmmm (4a,b)
  4 b 1̄     SL 10   0,1/2,0        Pmmm (1a-h)
  4 a 1̄     SL 10   0,0,0          Pmmm (1a-h)
```

Space group   No. 55   Pbam

Superlattices

```
SL  1   P(1/2,1,1)      density  2   add.gen.  1/2,0,0
SL  2   P(1,1/2,1)      density  2   add.gen.  0,1/2,0
SL  3   P(1,1,1/2)      density  2   add.gen.  0,0,1/2
SL  6   C(1,1,1)        density  2   add.gen.  1/2,1/2,0
SL  9   P(1/2,1/2,1)    density  4   add.gen.  1/2,0,0   0,1/2,0
SL 10   P(1/2,1,1/2)    density  4   add.gen.  1/2,0,0   0,0,1/2
SL 13   C(1/2,1,1)      density  4   add.gen.  1/4,1/2,0
SL 16   P(1,1/2,1/2)    density  4   add.gen.  0,1/2,0   0,0,1/2
SL 19   C(1,1/2,1)      density  4   add.gen.  1/2,1/4,0
SL 24   C(1,1,1/2)      density  4   add.gen.  0,0,1/2   1/2,1/2,0
SL 35   P(1/2,1/2,1/2)  density  8   add.gen.  1/2,0,0   0,1/2,0   0,0,1/2
SL 43   C(1/2,1,1/2)    density  8   add.gen.  0,0,1/2   1/4,1/2,0
SL 55   C(1,1/2,1/2)    density  8   add.gen.  0,0,1/2   1/2,1/4,0
```

Wyckoff letter   non-characteristic crystallographic orbits

```
 8 i 1      SL  1   x,1/4,z       Pbmm (4k)
            SL  2   1/4,y,z       Pmam (4k)
            SL  3   x,y,1/4       Pbam (4g,h)
            SL  6   x,0,z         Cmmm (8n,o)    0,y,z         Cmmm (8n,o)
            SL  9   1/4,0,z       Pmmm (2i-t)    0,1/4,z       Pmmm (2i-t)
                    1/4,1/4,z     Pmmm (2i-t)
            SL 10   x,1/4,1/4     Pbmm (2e,f)
            SL 13   3/8,1/4,z     Cmmm (4k,l)    1/8,1/4,z     Cmmm (4k,l)
            SL 16   1/4,y,1/4     Pmam (2e,f)
            SL 19   1/4,3/8,z     Cmmm (4k,l)    1/4,1/8,z     Cmmm (4k,l)
            SL 24   x,0,1/4       Cmmm (4g-j)    0,y,1/4       Cmmm (4g-j)
            SL 35   1/4,0,1/4     Pmmm (1a-h)    0,1/4,1/4     Pmmm (1a-h)
                    1/4,1/4,1/4   Pmmm (1a-h)
            SL 43   3/8,1/4,1/4   Cmmm (2a-d)    1/8,1/4,1/4   Cmmm (2a-d)
            SL 55   1/4,3/8,1/4   Cmmm (2a-d)    1/4,1/8,1/4   Cmmm (2a-d)
 4 h m      SL  1   x,1/4,1/2     Pbmm (2e,f)
            SL  2   1/4,y,1/2     Pmam (2e,f)
            SL  6   x,0,1/2       Cmmm (4g-j)    0,y,1/2       Cmmm (4g-j)
            SL  9   1/4,0,1/2     Pmmm (1a-h)    0,1/4,1/2     Pmmm (1a-h)
                    1/4,1/4,1/2   Pmmm (1a-h)
            SL 13   3/8,1/4,1/2   Cmmm (2a-d)    1/8,1/4,1/2   Cmmm (2a-d)
            SL 19   1/4,3/8,1/2   Cmmm (2a-d)    1/4,1/8,1/2   Cmmm (2a-d)
 4 g m      SL  1   x,1/4,0       Pbmm (2e,f)
            SL  2   1/4,y,0       Pmam (2e,f)
            SL  6   x,0,0         Cmmm (4g-j)    0,y,0         Cmmm (4g-j)
            SL  9   1/4,0,0       Pmmm (1a-h)    0,1/4,0       Pmmm (1a-h)
                    1/4,1/4,0     Pmmm (1a-h)
            SL 13   3/8,1/4,0     Cmmm (2a-d)    1/8,1/4,0     Cmmm (2a-d)
            SL 19   1/4,3/8,0     Cmmm (2a-d)    1/4,1/8,0     Cmmm (2a-d)
 4 f 2      SL  6   0,1/2,z       Cmmm (4k,l)
            SL 24   0,1/2,1/4     Cmmm (2a-d)
 4 e 2      SL  6   0,0,z         Cmmm (4k,l)
            SL 24   0,0,1/4       Cmmm (2a-d)
 2 d 2/m    SL  6   0,1/2,1/2     Cmmm (2a-d)
 2 c 2/m    SL  6   0,1/2,0       Cmmm (2a-d)
 2 b 2/m    SL  6   0,0,1/2       Cmmm (2a-d)
 2 a 2/m    SL  6   0,0,0         Cmmm (2a-d)
```

Space group No. 56 Pccn

Superlattices

SL	3	P(1,1,1/2)	density	2	add.gen.	0,0,1/2
SL	4	A(1,1,1)	density	2	add.gen.	0,1/2,1/2
SL	5	B(1,1,1)	density	2	add.gen.	1/2,0,1/2
SL	6	C(1,1,1)	density	2	add.gen.	1/2,1/2,0
SL	7	I(1,1,1)	density	2	add.gen.	1/2,1/2,1/2
SL	10	P(1/2,1,1/2)	density	4	add.gen.	1/2,0,0 0,0,1/2
SL	11	A(1/2,1,1)	density	4	add.gen.	1/2,0,0 0,1/2,1/2
SL	16	P(1,1/2,1/2)	density	4	add.gen.	0,1/2,0 0,0,1/2
SL	18	B(1,1/2,1)	density	4	add.gen.	0,1/2,0 1/2,0,1/2
SL	24	C(1,1,1/2)	density	4	add.gen.	0,0,1/2 1/2,1/2,0
SL	25	I(1,1,1/2)	density	4	add.gen.	1/2,1/2,1/4
SL	26	F(1,1,1)	density	4	add.gen.	0,1/2,1/2 1/2,0,1/2
SL	35	P(1/2,1/2,1/2)	density	8	add.gen.	1/2,0,0 0,1/2,0 0,0,1/2
SL	41	A(1/2,1,1/2)	density	8	add.gen.	1/2,0,0 0,1/2,1/4
SL	45	F(1/2,1,1)	density	8	add.gen.	0,1/2,1/2 1/4,0,1/2
SL	54	B(1,1/2,1/2)	density	8	add.gen.	0,1/2,0 1/2,0,1/4
SL	57	F(1,1/2,1)	density	8	add.gen.	0,1/4,1/2 1/2,0,1/2

Wyckoff letter   non-characteristic crystallographic orbits

8 e 1	SL 3	x,1/4,z	Pmmn (4e,f)	1/4,y,z	Pmmn (4e,f)
	SL 4	x,0,z	Abma (8f)		
	SL 5	0,y,z	Bmab (8f)		
	SL 6	x,y,0	Cccm (8l)		
	SL 7	x,y,1/4	Ibam (8j)		
	SL10	0,1/4,z	Pmmb (2e,f)		
	SL11	x,0,1/4	Abmm (4g)		
	SL16	1/4,0,z	Pmma (2e,f)		
	SL18	0,y,1/4	Bmam (4g)		
	SL24	x,1/4,0	Cmmm (4g-j)	1/4,y,1/4	Cmmm (4g-j)
		1/4,y,0	Cmmm (4g-j)	x,1/4,1/4	Cmmm (4g-j)
	SL25	x,3/4,1/8	Immm (4e-j)	x,1/4,1/8	Immm (4e-j)
		1/4,y,3/8	Immm (4e-j)	1/4,y,1/8	Immm (4e-j)
	SL26	x,0,0	Fmmm (8g-i)	0,y,0	Fmmm (8g-i)
		0,0,z	Fmmm (8g-i)		
	SL35	1/4,0,0	Pmmm (1a-h)	0,0,1/4	Pmmm (1a-h)
		1/4,0,1/4	Pmmm (1a-h)	0,1/4,0	Pmmm (1a-h)
		0,1/4,1/4	Pmmm (1a-h)		
	SL41	0,3/4,1/8	Ammm (2a-d)	0,1/4,1/8	Ammm (2a-d)
	SL45	1/8,1/2,1/4	Fmmm (4a,b)	1/8,0,1/4	Fmmm (4a,b)
	SL54	1/4,0,3/8	Bmmm (2a-d)	1/4,0,1/8	Bmmm (2a-d)
	SL57	0,3/8,1/4	Fmmm (4a,b)	0,1/8,1/4	Fmmm (4a,b)
4 d 2	SL 3	1/4,3/4,z	Pmmn (2a,b)		
	SL24	1/4,3/4,1/4	Cmmm (2a-d)	1/4,3/4,0	Cmmm (2a-d)
	SL25	1/4,3/4,3/8	Immm (2a-d)	1/4,3/4,1/8	Immm (2a-d)
4 c 2	SL 3	1/4,1/4,z	Pmmn (2a,b)		
	SL24	1/4,1/4,0	Cmmm (2a-d)	1/4,1/4,1/4	Cmmm (2a-d)
	SL25	1/4,1/4,3/8	Immm (2a-d)	1/4,1/4,1/8	Immm (2a-d)
4 b 1̄	SL26	0,0,1/2	Fmmm (4a,b)		
4 a 1̄	SL26	0,0,0	Fmmm (4a,b)		

Space group No. 57 Pbcm

Superlattices

```
SL  2  P(1,1/2,1)       density  2  add.gen.  0,1/2,0
SL  3  P(1,1,1/2)       density  2  add.gen.  0,0,1/2
SL  4  A(1,1,1)         density  2  add.gen.  0,1/2,1/2
SL  6  C(1,1,1)         density  2  add.gen.  1/2,1/2,0
SL 16  P(1,1/2,1/2)     density  4  add.gen.  0,1/2,0  0,0,1/2
SL 17  A(1,1/2,1)       density  4  add.gen.  0,1/4,1/2
SL 21  P(1,1,1/4)       density  4  add.gen.  0,0,1/4
SL 24  C(1,1,1/2)       density  4  add.gen.  0,0,1/2  1/2,1/2,0
SL 26  F(1,1,1)         density  4  add.gen.  0,1/2,1/2  1/2,0,1/2
SL 35  P(1/2,1/2,1/2)   density  8  add.gen.  1/2,0,0  0,1/2,0  0,0,1/2
SL 47  P(1,1/4,1/2)     density  8  add.gen.  0,1/4,0  0,0,1/2
SL 52  P(1,1/2,1/4)     density  8  add.gen.  0,1/2,0  0,0,1/4
SL 61  C(1,1,1/4)       density  8  add.gen.  0,0,1/4  1/2,1/2,0
```

Wyckoff letter   non-characteristic crystallographic orbits

```
8 e 1      SL  2  1/2,y,z        Pmcm (4i,j)    0,y,z          Pmcm (4i,j)
           SL  3  x,1/4,z        Pbmm (4k)      x,y,0          Pbmm (4i,j)
           SL  4  x,0,z          Abmm (8m,n)
           SL  6  1/4,y,z        Cmcm (8f)
           SL 16  1/2,1/4,z      Pmmm (2i-t)    0,1/4,z        Pmmm (2i-t)
                  1/2,y,0        Pmmm (2i-t)    0,y,0          Pmmm (2i-t)
                  1/2,0,z        Pmmm (2i-t)    0,0,z          Pmmm (2i-t)
                  x,0,0          Pmmm (2i-t)
           SL 17  1/2,1/8,z      Ammm (4g-j)    0,1/8,z        Ammm (4g-j)
           SL 21  x,1/4,1/8      Pbmm (2e,f)
           SL 24  1/4,3/4,z      Cmmm (4k,l)    1/4,1/4,z      Cmmm (4k,l)
                  1/4,y,0        Cmmm (4g-j)
           SL 26  1/4,0,z        Fmmm (8g-i)
           SL 35  1/4,0,0        Pmmm (1a-h)
           SL 47  1/2,1/8,0      Pmmm (1a-h)    0,1/8,0        Pmmm (1a-h)
           SL 52  1/2,1/4,1/8    Pmmm (1a-h)    0,1/4,1/8      Pmmm (1a-h)
                  1/2,0,1/8      Pmmm (1a-h)    0,0,1/8        Pmmm (1a-h)
           SL 61  1/4,3/4,1/8    Cmrm (2a-d)    1/4,1/4,1/8    Cmmm (2a-d)
4 d m      SL  2  1/2,y,1/4      Pmcm (2e,f)    0,y,1/4        Pmcm (2e,f)
           SL  3  x,1/4,1/4      Pbmm (2e,f)
           SL  4  x,0,1/4        Abmm (4g)
           SL  6  1/4,y,1/4      Cmcm (4c)
           SL 16  1/2,1/4,1/4    Pmmm (1a-h)    0,1/4,1/4      Pmmm (1a-h)
                  1/2,0,1/4      Pmmm (1a-h)    0,0,1/4        Pmmm (1a-h)
           SL 17  1/2,1/8,1/4    Ammm (2a-d)    1/2,3/8,1/4    Ammm (2a-d)
                  0,1/8,1/4      Ammm (2a-d)    0,3/8,1/4      Ammm (2a-d)
           SL 24  1/4,3/4,1/4    Cmmm (2a-d)    1/4,1/4,1/4    Cmmm (2a-d)
           SL 26  1/4,0,1/4      Fmmm (4a,b)    1/4,1/2,1/4    Fmmm (4a,b)
4 c 2      SL  3  x,1/4,0        Pbmm (2e,f)
           SL 16  1/2,1/4,0      Pmmm (1a-h)    0,1/4,0        Pmmm (1a-h)
           SL 24  3/4,1/4,0      Cmmm (2a-d)    1/4,1/4,0      Cmmm (2a-d)
4 b 1̄     SL 16  1/2,0,0        Pmmm (1a-h)
4 a 1̄     SL 16  0,0,0          Pmmm (1a-h)
```

Space group No. 58 Pnnm

Superlattices

```
SL  3  P(1,1,1/2)        density  2  add.gen.  0,0,1/2
SL  4  A(1,1,1)          density  2  add.gen.  0,1/2,1/2
SL  5  B(1,1,1)          density  2  add.gen.  1/2,0,1/2
SL  7  I(1,1,1)          density  2  add.gen.  1/2,1/2,1/2
SL 10  P(1/2,1,1/2)      density  4  add.gen.  1/2,0,0  0,0,1/2
SL 11  A(1/2,1,1)        density  4  add.gen.  1/2,0,0  0,1/2,1/2
SL 16  P(1,1/2,1/2)      density  4  add.gen.  0,1/2,0  0,0,1/2
SL 18  B(1,1/2,1)        density  4  add.gen.  0,1/2,0  1/2,0,1/2
SL 24  C(1,1,1/2)        density  4  add.gen.  0,0,1/2  1/2,1/2,0
SL 26  F(1,1,1)          density  4  add.gen.  0,1/2,1/2  1/2,0,1/2
SL 35  P(1/2,1/2,1/2)    density  8  add.gen.  1/2,0,0  0,1/2,0  0,0,1/2
SL 43  C(1/2,1,1/2)      density  8  add.gen.  0,0,1/2  1/4,1/2,0
SL 55  C(1,1/2,1/2)      density  8  add.gen.  0,0,1/2  1/2,1/4,0
```

Wyckoff letter   non-characteristic crystallographic orbits

```
8 h 1     SL  3  x,y,1/4         Pbam (4g,h)
          SL  4  1/4,y,z         Amam (8g)
          SL  5  x,1/4,z         Bbmm (8g)
          SL  7  x,0,z           Immm (8l-n)      0,y,z          Immm (8l-n)
          SL10  x,1/4,1/4        Pbmm (2e,f)
          SL11  1/4,0,z          Ammm (4g-j)
          SL16  1/4,y,1/4        Pmam (2e,f)
          SL18  0,1/4,z          Bmmm (4g-j)
          SL24  0,y,1/4          Cmmm (4g-j)      x,0,1/4        Cmmm (4g-j)
          SL26  1/4,1/4,z        Fmmm (8g-i)
          SL35  1/4,0,1/4        Pmmm (1a-h)      0,1/4,1/4      Pmmm (1a-h)
                1/4,1/4,1/4      Pmmm (1a-h)
          SL43  3/8,1/4,1/4      Cmmm (2a-d)      1/8,1/4,1/4    Cmmm (2a-d)
          SL55  1/4,3/8,1/4      Cmmm (2a-d)      1/4,1/8,1/4    Cmmm (2a-d)
4 g m     SL  4  1/4,y,0         Amam (4c)
          SL  5  x,1/4,0         Bbmm (4c)
          SL  7  x,1/2,0         Immm (4e-j)      x,0,0          Immm (4e-j)
                1/2,y,0          Immm (4e-j)      0,y,0          Immm (4e-j)
          SL11  1/4,1/2,0        Ammm (2a-d)      1/4,0,0        Ammm (2a-d)
          SL18  1/2,1/4,0        Bmmm (2a-d)      0,1/4,0        Bmmm (2a-d)
          SL26  1/4,3/4,0        Fmmm (4a,b)      1/4,1/4,0      Fmmm (4a,b)
4 f 2     SL  7  0,1/2,z         Immm (4e-j)
          SL24  0,1/2,1/4        Cmmm (2a-d)
4 e 2     SL  7  0,0,z           Immm (4e-j)
          SL24  0,0,1/4          Cmmm (2a-d)
2 d 2/m   SL  7  0,1/2,1/2       Immm (2a-d)
2 c 2/m   SL  7  0,1/2,0         Immm (2a-d)
2 b 2/m   SL  7  0,0,1/2         Immm (2a-d)
2 a 2/m   SL  7  0,0,0           Immm (2a-d)
```

Space group No. 59 Pmmn

Superlattices

```
SL   1   P(1/2,1,1)      density  2   add.gen.   1/2,0,0
SL   2   P(1,1/2,1)      density  2   add.gen.   0,1/2,0
SL   6   C(1,1,1)        density  2   add.gen.   1/2,1/2,0
SL   7   I(1,1,1)        density  2   add.gen.   1/2,1/2,1/2
SL   9   P(1/2,1/2,1)    density  4   add.gen.   1/2,0,0   0,1/2,0
SL  11   A(1/2,1,1)      density  4   add.gen.   1/2,0,0   0,1/2,1/2
SL  18   B(1,1/2,1)      density  4   add.gen.   0,1/2,0   1/2,0,1/2
SL  28   P(1/4,1/2,1)    density  8   add.gen.   1/4,0,0   0,1/2,0
SL  34   P(1/2,1/4,1)    density  8   add.gen.   1/2,0,0   0,1/4,0
SL  35   P(1/2,1/2,1/2)  density  8   add.gen.   1/2,0,0   0,1/2,0   0,0,1/2
```

Wyckoff letter   non-characteristic crystallographic orbits

```
    8 g 1     SL  1   0,y,z         Pmmb (4i,j)
              SL  2   x,0,z         Pmma (4i,j)
              SL  6   x,y,1/2       Cmmm (8p,q)     x,y,0          Cmmm (8p,q)
              SL  7   x,y,1/4       Immm (8l-n)
              SL  9   0,0,z         Pmmm (2i-t)     0,y,1/2        Pmmm (2i-t)
                      0,y,0         Pmmm (2i-t)     x,0,1/2        Pmmm (2i-t)
                      x,0,0         Pmmm (2i-t)
              SL 11   0,y,1/4       Ammm (4g-j)
              SL 18   x,0,1/4       Bmmm (4g-j)
              SL 28   1/8,0,1/2     Pmmm (1a-h)     1/8,0,0        Pmmm (1a-h)
              SL 34   0,1/8,1/2     Pmmm (1a-h)     0,1/8,0        Pmmm (1a-h)
              SL 35   0,0,1/4       Pmmm (1a-h)
    4 f m     SL  1   0,1/4,z       Pmmb (2e,f)
              SL  6   x,1/4,1/2     Cmmm (4g-j)     x,1/4,0        Cmmm (4g-j)
              SL  7   x,1/4,3/4     Immm (4e-j)     x,1/4,1/4      Immm (4e-j)
              SL  9   0,1/4,1/2     Pmmm (1a-h)     0,1/4,0        Pmmm (1a-h)
              SL 11   0,1/4,3/4     Ammm (2a-d)     0,1/4,1/4      Ammm (2a-d)
    4 e m     SL  2   1/4,0,z       Pmma (2e,f)
              SL  6   1/4,y,1/2     Cmmm (4g-j)     1/4,y,0        Cmmm (4g-j)
              SL  7   1/4,y,3/4     Immm (4e-j)     1/4,y,1/4      Immm (4e-j)
              SL  9   1/4,0,1/2     Pmmm (1a-h)     1/4,0,0        Pmmm (1a-h)
              SL 18   1/4,0,3/4     Bmmm (2a-d)     1/4,0,1/4      Bmmm (2a-d)
    4 d 1     SL  9   0,0,1/2       Pmmm (1a-h)
    4 c 1     SL  9   0,0,0         Pmmm (1a-h)
    2 b mm2   SL  6   1/4,3/4,1/2   Cmmm (2a-d)     1/4,3/4,0      Cmmm (2a-d)
              SL  7   1/4,3/4,1/4   Immm (2a-d)     1/4,3/4,3/4    Immm (2a-d)
    2 a mm2   SL  6   1/4,1/4,1/2   Cmmm (2a-d)     1/4,1/4,0      Cmmm (2a-d)
              SL  7   1/4,1/4,3/4   Immm (2a-d)     1/4,1/4,1/4    Immm (2a-d)
```

Space group No. 60 Pbcn

Superlattices

```
SL   2   P(1,1/2,1)        density  2  add.gen.  0,1/2,0
SL   3   P(1,1,1/2)        density  2  add.gen.  0,0,1/2
SL   4   A(1,1,1)          density  2  add.gen.  0,1/2,1/2
SL   6   C(1,1,1)          density  2  add.gen.  1/2,1/2,0
SL   7   I(1,1,1)          density  2  add.gen.  1/2,1/2,1/2
SL   9   P(1/2,1/2,1)      density  4  add.gen.  1/2,0,0   0,1/2,0
SL  11   A(1/2,1,1)        density  4  add.gen.  1/2,0,0   0,1/2,1/2
SL  16   P(1,1/2,1/2)      density  4  add.gen.  0,1/2,0   0,0,1/2
SL  17   A(1,1/2,1)        density  4  add.gen.  0,1/4,1/2
SL  18   B(1,1/2,1)        density  4  add.gen.  0,1/2,0   1/2,0,1/2
SL  24   C(1,1,1/2)        density  4  add.gen.  0,0,1/2   1/2,1/2,0
SL  25   I(1,1,1/2)        density  4  add.gen.  1/2,1/2,1/4
SL  26   F(1,1,1)          density  4  add.gen.  0,1/2,1/2   1/2,0,1/2
SL  35   P(1/2,1/2,1/2)    density  8  add.gen.  1/2,0,0   0,1/2,0   0,0,1/2
SL  36   A(1/2,1/2,1)      density  8  add.gen.  1/2,0,0   0,1/4,1/2
SL  45   F(1/2,1,1)        density  8  add.gen.  0,1/2,1/2   1/4,0,1/2
SL  54   B(1,1/2,1/2)      density  8  add.gen.  0,1/2,0   1/2,0,1/4
SL  57   F(1,1/2,1)        density  8  add.gen.  0,1/4,1/2   1/2,0,1/2
SL  61   C(1,1,1/4)        density  8  add.gen.  0,0,1/4   1/2,1/2,0
```

Wyckoff letter   non-characteristic crystallographic orbits

```
8 d 1     SL  2   1/4,y,z        Pmca (4d)
          SL  3   x,0,z          Pbmn (4h)
          SL  4   x,1/4,z        Abma (8f)
          SL  6   x,y,1/4        Cmcm (8g)        0,y,z          Cmcm (8f)
          SL  7   x,y,0          Ibam (8j)
          SL  9   1/4,y,1/4      Pmcm (2e,f)
          SL 11   x,1/4,0        Abmm (4g)
          SL 16   1/4,1/4,z      Pmma (2e,f)      1/4,0,z        Pmma (2e,f)
          SL 17   1/4,1/8,z      Amma (4c)
          SL 18   1/4,y,0        Bmcm (4g)
          SL 24   0,1/2,z        Cmmm (4k,l)      0,0,z          Cmmm (4k,l)
                  x,0,1/4        Cmmm (4g-j)      x,0,0          Cmmm (4g-j)
                  0,y,0          Cmmm (4g-j)
          SL 25   x,0,1/8        Ibmm (4e)
          SL 26   x,1/4,1/4      Fmmm (8g-i)      0,1/4,z        Fmmm (8g-i)
          SL 35   1/4,0,0        Pmmm (1a-h)      1/4,0,1/4      Pmmm (1a-h)
                  1/4,1/4,1/4    Pmmm (1a-h)      0,1/4,0        Pmmm (1a-h)
                  1/4,1/4,0      Pmmm (1a-h)
          SL 36   1/4,3/8,1/4    Ammm (2a-d)      1/4,1/8,1/4    Ammm (2a-d)
          SL 45   1/8,1/4,1/2    Fmmm (4a,b)      1/8,1/4,0      Fmmm (4a,b)
          SL 54   1/4,0,3/8      Bmmm (2a-d)      1/4,0,1/8      Bmmm (2a-d)
                  1/4,1/4,3/8    Bmmm (2a-d)      1/4,1/4,1/8    Bmmm (2a-d)
          SL 57   1/4,1/8,1/2    Fmmm (4a,b)      1/4,1/8,0      Fmmm (4a,b)
          SL 61   0,1/2,1/8      Cmmm (2a-d)      0,0,1/8        Cmmm (2a-d)
4 c 2     SL  6   0,y,1/4        Cmcm (4c)
          SL 24   0,1/2,1/4      Cmmm (2a-d)      0,0,1/4        Cmmm (2a-d)
          SL 26   0,3/4,1/4      Fmmm (4a,b)      0,1/4,1/4      Fmmm (4a,b)
4 b 1̄     SL 24   0,1/2,0        Cmmm (2a-d)
4 a 1̄     SL 24   0,0,0          Cmmm (2a-d)
```

Space group No. 61 Pbca

Superlattices

```
SL  1  P(1/2,1,1)      density  2  add.gen.  1/2,0,0
SL  2  P(1,1/2,1)      density  2  add.gen.  0,1/2,0
SL  3  P(1,1,1/2)      density  2  add.gen.  0,0,1/2
SL  4  A(1,1,1)        density  2  add.gen.  0,1/2,1/2
SL  5  B(1,1,1)        density  2  add.gen.  1/2,0,1/2
SL  6  C(1,1,1)        density  2  add.gen.  1/2,1/2,0
SL  9  P(1/2,1/2,1)    density  4  add.gen.  1/2,0,0   0,1/2,0
SL 10  P(1/2,1,1/2)    density  4  add.gen.  1/2,0,0   0,0,1/2
SL 11  A(1/2,1,1)      density  4  add.gen.  1/2,0,0   0,1/2,1/2
SL 13  C(1/2,1,1)      density  4  add.gen.  1/4,1/2,0
SL 16  P(1,1/2,1/2)    density  4  add.gen.  0,1/2,0   0,0,1/2
SL 17  A(1,1/2,1)      density  4  add.gen.  0,1/4,1/2
SL 18  B(1,1/2,1)      density  4  add.gen.  0,1/2,0   1/2,0,1/2
SL 23  B(1,1,1/2)      density  4  add.gen.  1/2,0,1/4
SL 24  C(1,1,1/2)      density  4  add.gen.  0,0,1/2   1/2,1/2,0
SL 26  F(1,1,1)        density  4  add.gen.  0,1/2,1/2  1/2,0,1/2
SL 35  P(1/2,1/2,1/2)  density  8  add.gen.  1/2,0,0   0,1/2,0  0,0,1/2
SL 36  A(1/2,1/2,1)    density  8  add.gen.  1/2,0,0   0,1/4,1/2
SL 43  C(1/2,1,1/2)    density  8  add.gen.  0,0,1/2   1/4,1/2,0
SL 45  F(1/2,1,1)      density  8  add.gen.  0,1/2,1/2  1/4,0,1/2
SL 54  B(1,1/2,1/2)    density  8  add.gen.  0,1/2,0   1/2,0,1/4
SL 57  F(1,1/2,1)      density  8  add.gen.  0,1/4,1/2  1/2,0,1/2
SL 63  F(1,1,1/2)      density  8  add.gen.  0,1/2,1/4  1/2,1/2,0
```

Wyckoff letter   non-characteristic crystallographic orbits

```
8 c 1     SL  1   x,y,1/4       Pbcm (4d)
          SL  2   1/4,y,z       Pmca (4d)
          SL  3   x,1/4,z       Pbma (4d)
          SL  4   x,0,z         Abma (8f)
          SL  5   x,y,0         Bbcm (8f)
          SL  6   0,y,z         Cmca (8f)
          SL  9   0,y,1/4       Pmcm (2e,f)     1/4,y,1/4      Pmcm (2e,f)
          SL 10   x,1/4,1/4     Pbmm (2e,f)     x,1/4,0        Pbmm (2e,f)
          SL 11   x,0,1/4       Abmm (4g)
          SL 13   1/8,y,1/4     Cmcm (4c)
          SL 16   1/4,0,z       Pmma (2e,f)     1/4,1/4,z      Pmma (2e,f)
          SL 17   1/4,1/8,z     Amma (4c)
          SL 18   1/4,y,0       Bmcm (4g)
          SL 23   x,1/4,1/8     Bbmm (4c)
          SL 24   0,1/4,z       Cmma (4g)
          SL 26   0,y,0         Fmmm (8g-i)     x,0,0          Fmmm (8g-i)
                  0,0,z         Fmmm (8g-i)
          SL 35   1/4,0,0       Pmmm (1a-h)     1/4,1/4,1/4    Pmmm (1a-h)
                  1/4,0,1/4     Pmmm (1a-h)     0,0,1/4        Pmmm (1a-h)
                  0,1/4,1/4     Pmmm (1a-h)     0,1/4,0        Pmmm (1a-h)
                  1/4,1/4,0     Pmmm (1a-h)
          SL 36   1/4,3/8,1/4   Ammm (2a-d)     1/4,1/8,1/4    Ammm (2a-d)
                  0,3/8,1/4     Ammm (2a-d)     0,1/8,1/4      Ammm (2a-d)
          SL 43   3/8,1/4,1/4   Cmmm (2a-d)     1/8,1/4,1/4    Cmmm (2a-d)
                  3/8,1/4,0     Cmmm (2a-d)     1/8,1/4,0      Cmmm (2a-d)
          SL 45   1/8,1/2,1/4   Fmmm (4a,b)     1/8,0,1/4      Fmmm (4a,b)
```

(continued on next page)

Space group No. 61 Pbca   (continued)

Wyckoff letter   non-characteristic crystallographic orbits

8 c 1	SL54	1/4,1/4,3/8	Bmmm (2a-d)		1/4,1/4,1/8	Bmmm (2a-d)
		1/4,0,3/8	Bmmm (2a-d)		1/4,0,1/8	Bmmm (2a-d)
	SL57	1/4,1/8,1/2	Fmmm (4a,b)		1/4,1/8,0	Fmmm (4a,b)
	SL63	1/2,1/4,1/8	Fmmm (4a,b)		0,1/4,1/8	Fmmm (4a,b)
4 b $\bar{1}$	SL26	0,0,1/2	Fmmm (4a,b)			
4 a $\bar{1}$	SL26	0,0,0	Fmmm (4a,b)			

Space group No. 62 Pnma

Superlattices

```
SL  1   P(1/2,1,1)        density  2  add.gen.  1/2,0,0
SL  2   P(1,1/2,1)        density  2  add.gen.  0,1/2,0
SL  4   A(1,1,1)          density  2  add.gen.  0,1/2,1/2
SL  5   B(1,1,1)          density  2  add.gen.  1/2,0.1/2
SL  7   I(1,1,1)          density  2  add.gen.  1/2,1/2,1/2
SL  9   P(1/2,1/2,1)      density  4  add.gen.  1/2,0,0   0,1/2,0
SL 11   A(1/2,1,1)        density  4  add.gen.  1/2,0,0   0,1/2,1/2
SL 14   I(1/2,1,1)        density  4  add.gen.  1/4,1/2,1/2
SL 16   P(1,1/2,1/2)      density  4  add.gen.  0,1/2,0   0,0,1/2
SL 18   B(1,1/2,1)        density  4  add.gen.  0,1/2,0   1/2,0,1/2
SL 26   F(1,1,1)          density  4  add.gen.  0,1/2,1/2  1/2,0,1/2
SL 35   P(1/2,1/2,1/2)    density  8  add.gen.  1/2,0,0   0,1/2,0   0,0,1/2
SL 37   B(1/2,1/2,1)      density  8  add.gen.  0,1/2,0   1/4,0,1/2
SL 49   B(1,1/4,1)'       density  8  add.gen.  0,1/4,0   1/2,0,1/2
SL 54   B(1,1/2,1/2)      density  8  add.gen.  0,1/2,0   1/2,0,1/4
```

Wyckoff letter   non-characteristic crystallographic orbits

Wyckoff	SL				
8 d 1	SL 1	x,y,1/4	Pnmm (4e,f)		
	SL 2	x,0,z	Pcma (4g,h)		
	SL 4	1/4,y,z	Amma (8g)		
	SL 5	x,y,0	Bbmm (8f)		
	SL 7	0,y,z	Imma (8h,i)		
	SL 9	x,0,1/4	Pcmm (2e,f)		
	SL11	1/4,y,1/4	Ammm (4g-j)	0,y,1/4	Ammm (4g-j)
	SL14	3/8,y,1/4	Immm (4e-j)	1/8,y,1/4	Immm (4e-j)
	SL16	1/4,0,z	Pmma (2e,f)		
	SL18	1/2,y,0	Bmmm (4k,l)	0,y,0	Bmmm (4k,l)
		x,0,0	Bmmm (4g-j)	0,0,z	Bmmm (4g-j)
	SL26	1/4,y,0	Fmmm (8g-i)		
	SL35	1/4,0,0	Pmmm (1a-h)	0,0,1/4	Pmmm (1a-h)
		1/4,0,1/4	Pmmm (1a-h)		
	SL37	3/8,0,1/4	Bmmm (2a-d)	1/8,0,1/4	Bmmm (2a-d)
	SL49	1/2,1/8,0	Bmmm (2a-d)	0,1/8,0	Bmmm (2a-d)
	SL54	3/4,0,1/8	Bmmm (2a-d)	1/4,0,1/8	Bmmm (2a-d)
4 c m	SL 1	x,1/4,3/4	Pnmm (2a,b)	x,1/4,1/4	Pnmm (2a,b)
	SL 4	1/4,1/4,z	Amma (4c)		
	SL 5	x,1/4,0	Bbmm (4c)		
	SL 7	0,1/4,z	Imma (4e)		
	SL11	1/4,1/4,3/4	Ammm (2a-d)	1/4,1/4,1/4	Ammm (2a-d)
		0,1/4,3/4	Ammm (2a-d)	0,1/4,1/4	Ammm (2a-d)
	SL14	1/8,1/4,3/4	Immm (2a-d)	3/8,1/4,1/4	Immm (2a-d)
		3/8,1/4,3/4	Immm (2a-d)	1/8,1/4,1/4	Immm (2a-d)
	SL18	1/2,1/4,0	Bmmm (2a-d)	0,1/4,0	Bmmm (2a-d)
	SL26	3/4,1/4,0	Fmmm (4a,b)	1/4,1/4,0	Fmmm (4a,b)
4 b 1̄	SL18	0,0,1/2	Bmmm (2a-d)		
4 a 1̄	SL18	0,0,0	Bmmm (2a-d)		

Space group No. 63 Cmcm

Superlattices

```
SL  1  P(1/2,1/2,1)    density  2  add.gen.  1/2,0,0
SL  2  C(1,1,1/2)      density  2  add.gen.  0,0,1/2
SL  3  F(1,1,1)        density  2  add.gen.  0,1/2,1/2
SL  6  P(1/2,1/2,1/2)  density  4  add.gen.  1/2,0,0    0,0,1/2
SL  7  A(1/2,1/2,1)    density  4  add.gen.  0,1/4,1/2
SL 11  C(1,1,1/4)      density  4  add.gen.  0,0,1/4
SL 15  P(1/4,1/2,1/2)  density  8  add.gen.  1/4,0,0    0,0,1/2
SL 21  P(1/2,1/4,1/2)  density  8  add.gen.  0,1/4,0    0,0,1/2
SL 26  P(1/2,1/2,1/4)  density  8  add.gen.  1/2,0,0    0,0,1/4
```

Wyckoff letter   non-characteristic crystallographic orbits

```
16 h 1    SL 1   1/4,y,z        Pmcm (4i,j)
          SL 2   x,y,0          Cmmm (8p,q)      x,0,z          Cmmm (8n,o)
          SL 3   x,1/4,z        Fmmm (16m-o)
          SL 6   x,1/4,0        Pmmm (2i-t)      1/4,1/4,z      Pmmm (2i-t)
                 1/4,0,z        Pmmm (2l-t)      1/4,y,0        Pmmm (2i-t)
          SL 7   1/4,1/8,z      Ammm (4g-j)
          SL11   x,0,1/8        Cmmm (4g-j)
          SL15   1/8,1/4,0      Pmmm (1a-h)
          SL21   1/4,1/8,0      Pmmm (1a-h)
          SL26   1/4,1/4,1/8    Pmmm (1a-h)      1/4,0,1/8      Pmmm (1a-h)
 8 g m    SL 1   1/4,y,1/4      Pmcm (2e,f)
          SL 2   x,0,1/4        Cmmm (4g-j)
          SL 3   x,1/4,1/4      Fmmm (8g-i)
          SL 6   1/4,1/4,1/4    Pmmm (1a-h)      1/4,0,1/4      Pmmm (1a-h)
          SL 7   1/4,1/8,1/4    Ammm (2a-d)      1/4,7/8,1/4    Ammm (2a-d)
 8 f m    SL 2   0,y,0          Cmmm (4g-j)      0,1/2,z        Cmmm (4k,l)
                 0,0,z          Cmmm (4k,l)
          SL 3   0,1/4,z        Fmmm (8g-i)
          SL 6   0,1/4,0        Pmmm (1a-h)
          SL11   0,1/2,1/8      Cmmm (2a-d)      0,0,1/8        Cmmm (2a-d)
 8 e 2    SL 2   x,0,0          Cmmm (4g-j)
          SL 6   1/4,0,0        Pmmm (1a-h)
 8 d 1̄    SL 6   1/4,1/4,0      Pmmm (1a-h)
 4 c mm2  SL 2   0,1/2,1/4      Cmmm (2a-d)      0,0,1/4        Cmmm (2a-d)
          SL 3   0,3/4,1/4      Fmmm (4a,b)      0,1/4,1/4      Fmmm (4a,b)
 4 b 2/m  SL 2   0,1/2,0        Cmmm (2a-d)
 4 a 2/m  SL 2   0,0,0          Cmmm (2a-d)
```

Space group No. 64 Cmca

Superlattices

SL	1	P(1/2,1/2,1)	density	2	add.gen.	1/2,0,0	
SL	2	C(1,1,1/2)	density	2	add.gen.	0,0,1/2	
SL	3	F(1,1,1)	density	2	add.gen.	0,1/2,1/2	
SL	4	P(1/4,1/2,1)	density	4	add.gen.	1/4,0,0	
SL	6	P(1/2,1/2,1/2)	density	4	add.gen.	1/2,0,0	0,0,1/2
SL	7	A(1/2,1/2,1)	density	4	add.gen.	0,1/4,1/2	
SL	12	F(1,1,1/2)	density	4	add.gen.	0,1/2,1/4	
SL	15	P(1/4,1/2,1/2)	density	8	add.gen.	1/4,0,0	0,0,1/2
SL	16	A(1/4,1/2,1)	density	8	add.gen.	1/4,0,0	0,1/4,1/2
SL	21	P(1/2,1/4,1/2)	density	8	add.gen.	0,1/4,0	0,0,1/2
SL	26	P(1/2,1/2,1/4)	density	8	add.gen.	1/2,0,0	0,0,1/4

Wyckoff letter		non-characteristic crystallographic orbits			
16 g 1	SL 1	x,y,1/4	Pmcm (4k)	1/4,y,z	Pmcm (4i,j)
	SL 2	x,1/4,z	Cmma (8m,n)		
	SL 3	x,0,z	Fmmm (16m-o)	x,y,0	Fmmm (16m-o)
	SL 4	1/8,y,1/4	Pmcm (2e,f)		
	SL 6	x,0,1/4	Pmmm (2i-t)	x,1/4,1/4	Pmmm (2i-t)
		1/4,y,0	Pmmm (2i-t)	1/4,0,z	Pmmm (2i-t)
		1/4,1/4,z	Pmmm (2i-t)	x,1/4,0	Pmmm (2i-t)
	SL 7	1/4,1/8,z	Ammm (4g-j)	x,3/8,1/4	Ammm (4k,l)
		x,1/8,1/4	Ammm (4k,l)		
	SL12	x,1/4,1/8	Fmmm (8g-i)		
	SL15	1/8,0,1/4	Pmmm (1a-h)	1/8,1/4,1/4	Pmmm (1a-h)
		1/8,1/4,0	Pmmm (1a-h)		
	SL16	1/8,1/8,1/4	Ammm (2a-d)	1/8,1/8,3/4	Ammm (2a-d)
	SL21	1/4,1/8,0	Pmmm (1a-h)		
	SL26	1/4,0,1/8	Pmmm (1a-h)	1/4,1/4,1/8	Pmmm (1a-h)
8 f m	SL 1	0,y,1/4	Pmcm (2e,f)		
	SL 2	0,1/4,z	Cmma (4g)		
	SL 3	0,0,z	Fmmm (8g-i)	0,y,0	Fmmm (8g-i)
	SL 6	0,0,1/4	Pmmm (1a-h)	0,1/4,1/4	Pmmm (1a-h)
		0,1/4,0	Pmmm (1a-h)		
	SL 7	0,3/8,1/4	Ammm (2a-d)	0,1/8,1/4	Ammm (2a-d)
	SL12	0,3/4,1/8	Fmmm (4a,b)	0,1/4,1/8	Fmmm (4a,b)
8 e 2	SL 1	1/4,y,1/4	Pmcm (2e,f)		
	SL 6	1/4,0,1/4	Pmmm (1a-h)	1/4,1/4,1/4	Pmmm (1a-h)
	SL 7	1/4,3/8,1/4	Ammm (2a-d)	1/4,1/8,1/4	Ammm (2a-d)
8 d 2	SL 3	x,0,0	Fmmm (8g-i)		
	SL 6	1/4,0,0	Pmmm (1a-h)		
8 c 1̄	SL 6	1/4,1/4,0	Pmmm (1a-h)		
4 b 2/m	SL 3	1/2,0,0	Fmmm (4a,b)		
4 a 2/m	SL 3	0,0,0	Fmmm (4a,b)		

Space group No. 65 Cmmm

Superlattices

```
SL   1   P(1/2,1/2,1)     density   2   add.gen.   1/2,0,0
SL   2   C(1,1,1/2)       density   2   add.gen.   0,0,1/2
SL   4   P(1/4,1/2,1)     density   4   add.gen.   1/4,0,0
SL   5   P(1/2,1/4,1)     density   4   add.gen.   0,1/4,0
SL   6   P(1/2,1/2,1/2)   density   4   add.gen.   1/2,0,0   0,0,1/2
SL  15   P(1/4,1/2,1/2)   density   8   add.gen.   1/4,0,0   0,0,1/2
SL  21   P(1/2,1/4,1/2)   density   8   add.gen.   0,1/4,0   0,0,1/2
```

Wyckoff letter   non-characteristic crystallographic orbits

```
16 r 1      SL  1   x,1/4,z          Pmmm (4u-z)    1/4,y,z      Pmmm (4u-z)
            SL  2   x,y,1/4          Cmmm (8p,q)
            SL  4   1/8,1/4,z        Pmmm (2i-t)
            SL  5   1/4,1/8,z        Pmmm (2i-t)
            SL  6   x,1/4,1/4        Pmmm (2i-t)    1/4,y,1/4    Pmmm (2i-t)
            SL 15   1/8,1/4,1/4      Pmmm (1a-h)
            SL 21   1/4,1/8,1/4      Pmmm (1a-h)
 8 q m      SL  1   x,1/4,1/2        Pmmm (2i-t)    1/4,y,1/2    Pmmm (2i-t)
            SL  4   1/8,1/4,1/2      Pmmm (1a-h)
            SL  5   1/4,1/8,1/2      Pmmm (1a-h)
 8 p m      SL  1   x,1/4,0          Pmmm (2i-t)    1/4,y,0      Pmmm (2i-t)
            SL  4   1/8,1/4,0        Pmmm (1a-h)
            SL  5   1/4,1/8,0        Pmmm (1a-h)
 8 o m      SL  1   1/4,0,z          Pmmm (2i-t)
            SL  2   x,0,1/4          Cmmm (4g-j)
            SL  6   1/4,0,1/4        Pmmm (1a-h)
 8 n m      SL  1   0,1/4,z          Pmmm (2i-t)
            SL  2   0,y,1/4          Cmmm (4g-j)
            SL  6   0,1/4,1/4        Pmmm (1a-h)
 8 m 2      SL  1   1/4,1/4,z        Pmmm (2i-t)
            SL  6   1/4,1/4,1/4      Pmmm (1a-h)
 4 l mm2    SL  2   0,1/2,1/4        Cmmm (2a-d)
 4 k mm2    SL  2   0,0,1/4          Cmmm (2a-d)
 4 j mm2    SL  1   0,1/4,1/2        Pmmm (1a-h)
 4 i mm2    SL  1   0,1/4,0          Pmmm (1a-h)
 4 h mm2    SL  1   1/4,0,1/2        Pmmm (1a-h)
 4 g mm2    SL  1   1/4,0,0          Pmmm (1a-h)
 4 f 2/m    SL  1   1/4,1/4,1/2      Pmmm (1a-h)
 4 e 2/m    SL  1   1/4,1/4,0        Pmmm (1a-h)
```

Space group   No. 66   Cccm

Superlattices

```
SL  2  C(1,1,1/2)       density  2  add.gen.  0,0,1/2
SL  3  F(1,1,1)         density  2  add.gen.  0,1/2,1/2
SL  6  P(1/2,1/2,1/2)   density  4  add.gen.  1/2,0,0   0,0,1/2
SL 11  C(1,1,1/4)       density  4  add.gen.  0,0,1/4
SL 15  P(1/4,1/2,1/2)   density  8  add.gen.  1/4,0,0   0,0,1/2
SL 21  P(1/2,1/4,1/2)   density  8  add.gen.  0,1/4,0   0,0,1/2
SL 26  P(1/2,1/2,1/4)   density  8  add.gen.  1/2,0,0   0,0,1/4
```

Wyckoff letter   non-characteristic crystallographic orbits

```
16 m 1      SL 2   x,y,1/4          Cmmm (8p,q)      0,y,z            Cmmm (8n,o)
   .               x,0,z            Cmmm (8n,o)
            SL 3   1/4,y,z          Fmmm (16m-o)     x,1/4,z          Fmmm (16m-o)
            SL 6   x,1/4,1/4        Pmmm (2i-t)      0,1/4,z          Pmmm (2i-t)
                   1/4,y,1/4        Pmmm (2i-t)      1/4,0,z          Pmmm (2i-t)
            SL11   x,0,1/8          Cmmm (4g-j)      0,y,1/8          Cmmm (4g-j)
            SL15   1/8,1/4,1/4      Pmmm (1a-h)
            SL21   1/4,1/8,1/4      Pmmm (1a-h)
            SL26   0,1/4,1/8        Pmmm (1a-h)      1/4,0,1/8        Pmmm (1a-h)
 8 l m      SL 2   0,y,0            Cmmm (4g-j)      x,0,0            Cmmm (4g-j)
            SL 3   1/4,y,0          Fmmm (8g-i)      x,1/4,0          Fmmm (8g-i)
            SL 6   0,1/4,0          Pmmm (1a-h)      1/4,0,0          Pmmm (1a-h)
 8 k 2      SL 3   1/4,1/4,z        Fmmm (8g-i)
            SL 6   1/4,1/4,1/4      Pmmm (1a-h)
 8 j 2      SL 2   0,1/2,z          Cmmm (4k,l)
            SL11   0,1/2,1/8        Cmmm (2a-d)
 8 i 2      SL 2   0,0,z            Cmmm (4k,l)
            SL11   0,0,1/8          Cmmm (2a-d)
 8 h 2      SL 2   0,y,1/4          Cmmm (4g-j)
            SL 6   0,1/4,1/4        Pmmm (1a-h)
 8 g 2      SL 2   x,0,1/4          Cmmm (4g-j)
            SL 6   1/4,0,1/4        Pmmm (1a-h)
 4 f 2/m    SL 3   1/4,3/4,0        Fmmm (4a,b)
 4 e 2/m    SL 3   1/4,1/4,0        Fmmm (4a,b)
 4 d 2/m    SL 2   0,1/2,0          Cmmm (2a-d)
 4 c 2/m    SL 2   0,0,0            Cmmm (2a-d)
 4 b 222    SL 2   0,1/2,1/4        Cmmm (2a-d)
 4 a 222    SL 2   0,0,1/4          Cmmm (2a-d)
```

Space group No. 67 Cmma

Superlattices

```
SL  1  P(1/2,1/2,1)    density 2  add.gen.  1/2,0,0
SL  3  F(1,1,1)        density 2  add.gen.  0,1/2,1/2
SL  4  P(1/4,1/2,1)    density 4  add.gen.  1/4,0,0
SL  5  P(1/2,1/4,1)    density 4  add.gen.  0,1/4,0
SL  6  P(1/2,1/2,1/2)  density 4  add.gen.  1/2,0,0  0,0,1/2
SL 14  P(1/4,1/4,1)    density 8  add.gen.  1/4,0,0  0,1/4,0
SL 15  P(1/4,1/2,1/2)  density 8  add.gen.  1/4,0,0  0,0,1/2
SL 21  P(1/2,1/4,1/2)  density 8  add.gen.  0,1/4,0  0,0,1/2
```

Wyckoff letter   non-characteristic crystallographic orbits

```
16 o 1    SL 1   1/4,y,z       Pmmm (4u-z)    x,0,z       Pmmm (4u-z)
                 x,y,1/2       Pmmm (4u-z)    x,y,0       Pmmm (4u-z)
          SL 3   x,y,1/4       Fmmm (16m-o)
          SL 4   1/8,y,1/2     Pmmm (2i-t)    1/8,y,0     Pmmm (2i-t)
                 1/8,0,z       Pmmm (2i-t)
          SL 5   x,1/8,1/2     Pmmm (2i-t)    x,1/8,0     Pmmm (2i-t)
                 1/4,1/8,z     Pmmm (2i-t)
          SL 6   1/4,y,1/4     Pmmm (2i-t)    x,0,1/4     Pmmm (2i-t)
          SL14   1/8,1/8,1/2   Pmmm (1a-h)    1/8,1/8,0   Pmmm (1a-h)
          SL15   1/8,0,1/4     Pmmm (1a-h)
          SL21   1/4,1/8,1/4   Pmmm (1a-h)
 8 n m    SL 1   1/4,1/4,z     Pmmm (2i-t)    x,1/4,1/2   Pmmm (2i-t)
                 x,1/4,0       Pmmm (2i-t)
          SL 3   x,1/4,1/4     Fmmm (8g-i)
          SL 4   1/8,1/4,1/2   Pmmm (1a-h)    1/8,1/4,0   Pmmm (1a-h)
          SL 6   1/4,1/4,1/4   Pmmm (1a-h)
 8 m m    SL 1   0,0,z         Pmmm (2i-t)    0,y,1/2     Pmmm (2i-t)
                 0,y,0         Pmmm (2i-t)
          SL 3   0,y,1/4       Fmmm (8g-i)
          SL 5   0,1/8,1/2     Pmmm (1a-h)    0,1/8,0     Pmmm (1a-h)
          SL 6   0,0,1/4       Pmmm (1a-h)
 8 l 2    SL 1   1/4,0,z       Pmmm (2i-t)
          SL 6   1/4,0,1/4     Pmmm (1a-h)
 8 k 2    SL 1   1/4,y,1/2     Pmmm (2i-t)
          SL 5   1/4,1/8,1/2   Pmmm (1a-h)
 8 j 2    SL 1   1/4,y,0       Pmmm (2i-t)
          SL 5   1/4,1/8,0     Pmmm (1a-h)
 8 i 2    SL 1   x,0,1/2       Pmmm (2i-t)
          SL 4   1/8,0,1/2     Pmmm (1a-h)
 8 h 2    SL 1   x,0,0         Pmmm (2i-t)
          SL 4   1/8,0,0       Pmmm (1a-h)
 4 g mm2  SL 1   0,1/4,1/2     Pmmm (1a-h)    0,1/4,0     Pmmm (1a-h)
          SL 3   0,1/4,1/4     Fmmm (4a,b)    0,1/4,3/4   Fmmm (4a,b)
 4 f 2/m  SL 1   1/4,1/4,1/2   Pmmm (1a-h)
 4 e 2/m  SL 1   1/4,1/4,0     Pmmm (1a-h)
 4 d 2/m  SL 1   0,0,1/2       Pmmm (1a-h)
 4 c 2/m  SL 1   0,0,0         Pmmm (1a-h)
 4 b 222  SL 1   1/4,0,1/2     Pmmm (1a-h)
 4 a 222  SL 1   1/4,0,0       Pmmm (1a-h)
```

Space group   No. 68   Ccca

Superlattices

SL	1	P(1/2,1/2,1)	density	2	add.gen.	1/2,0,0	
SL	2	C(1,1,1/2)	density	2	add.gen.	0,0,1/2	
SL	3	F(1,1,1)	density	2	add.gen.	0,1/2,1/2	
SL	6	P(1/2,1/2,1/2)	density	4	add.gen.	1/2,0,0 0,0,1/2	
SL	7	A(1/2,1/2,1)	density	4	add.gen.	0,1/4,1/2	
SL	8	B(1/2,1/2,1)	density	4	add.gen.	1/4,0,1/2	
SL	12	F(1,1,1/2)	density	4	add.gen.	0,1/2,1/4	
SL	15	P(1/4,1/2,1/2)	density	8	add.gen.	1/4,0,0 0,0,1/2	
SL	21	P(1/2,1/4,1/2)	density	8	add.gen.	0,1/4,0 0,0,1/2	
SL	26	P(1/2,1/2,1/4)	density	8	add.gen.	1/2,0,0 0,0,1/4	
SL	31	F(1/2,1/2,1)	density	8	add.gen.	0,1/4,1/2 1/4,1/4,0	

Wyckoff letter   non-characteristic crystallographic orbits

Wyckoff	SL	coord	group (orbit)	coord	group (orbit)
16 i 1	SL 1	x,y,0	Pccm (4q)		
	SL 2	1/4,y,z	Cmma (8m,n)	x,0,z	Cmma (8m,n)
	SL 3	x,y,1/4	Fmmm (16m-o)	x,1/4,z	Fmmm (16m-o)
		0,y,z	Fmmm (16m-o)		
	SL 6	0,y,0	Pmmm (2i-t)	x,0,0	Pmmm (2i-t)
		x,1/4,0	Pmmm (2i-t)	1/4,y,0	Pmmm (2i-t)
		1/4,y,1/4	Pmmm (2i-t)	0,0,z	Pmmm (2i-t)
		1/4,1/4,z	Pmmm (2i-t)	x,0,1/4	Pmmm (2i-t)
	SL 7	x,1/8,0	Acmm (4g)		
	SL 8	1/8,y,0	Bmcm (4g)		
	SL 12	x,0,1/8	Fmmm (8g-i)	1/4,y,1/8	Fmmm (8g-i)
	SL 15	1/8,0,0	Pmmm (1a-h)	1/8,0,1/4	Pmmm (1a-h)
		1/8,1/4,0	Pmmm (1a-h)		
	SL 21	1/4,1/8,1/4	Pmmm (1a-h)	1/4,1/8,0	Pmmm (1a-h)
		0,1/8,0	Pmmm (1a-h)		
	SL 26	0,0,1/8	Pmmm (1a-h)	1/4,1/4,1/8	Pmmm (1a-h)
	SL 31	1/8,1/8,0	Fmmm (4a,b)	1/8,1/8,1/2	Fmmm (4a,b)
8 h 2	SL 2	1/4,0,z	Cmma (4g)		
	SL 6	1/4,0,0	Pmmm (1a-h)	1/4,0,1/4	Pmmm (1a-h)
	SL 12	1/4,0,1/8	Fmmm (4a,b)	1/4,0,3/8	Fmmm (4a,b)
8 g 2	SL 3	0,1/4,z	Fmmm (8g-i)		
	SL 6	0,1/4,0	Pmmm (1a-h)		
8 f 2	SL 3	0,y,1/4	Fmmm (8g-i)		
	SL 6	0,0,1/4	Pmmm (1a-h)		
8 e 2	SL 3	x,1/4,1/4	Fmmm (8g-i)		
	SL 6	1/4,1/4,1/4	Pmmm (1a-h)		
8 d 1̄	SL 6	0,0,0	Pmmm (1a-h)		
8 c 1̄	SL 6	1/4,3/4,0	Pmmm (1a-h)		
4 b 222	SL 3	0,1/4,3/4	Fmmm (4a,b)		
4 a 222	SL 3	0,1/4,1/4	Fmmm (4a,b)		

Space group   No. 69   Fmmm

Superlattices

```
SL  1   P(1/2,1/2,1/2)   density 2   add.gen.   1/2,0,0
SL  2   P(1/4,1/2,1/2)   density 4   add.gen.   1/4,0,0
SL  3   P(1/2,1/4,1/2)   density 4   add.gen.   0,1/4,0
SL  4   P(1/2,1/2,1/4)   density 4   add.gen.   0,0,1/4
SL 10   P(1/4,1/4,1/2)   density 8   add.gen.   1/4,0,0   0,1/4,0
SL 11   P(1/4,1/2,1/4)   density 8   add.gen.   1/4,0,0   0,0,1/4
SL 17   P(1/2,1/4,1/4)   density 8   add.gen.   0,1/4,0   0,0,1/4
```

Wyckoff letter   non-characteristic crystallographic orbits

```
32 p 1      SL  1   1/4,y,z       Pmmm (4u-z)   x,y,1/4       Pmmm (4u-z)
                    x,1/4,z       Pmmm (4u-z)
            SL  2   1/8,y,1/4     Pmmm (2i-t)   1/8,1/4,z     Pmmm (2i-t)
            SL  3   x,1/8,1/4     Pmmm (2i-t)   1/4,1/8,z     Pmmm (2i-t)
            SL  4   x,1/4,1/8     Pmmm (2i-t)   1/4,y,1/8     Pmmm (2i-t)
            SL10    1/8,1/8,1/4   Pmmm (1a-h)
            SL11    1/8,1/4,1/8   Pmmm (1a-h)
            SL17    1/4,1/8,1/8   Pmmm (1a-h)
16 o m      SL  1   1/4,y,0       Pmmm (2i-t)   x,1/4,0       Pmmm (2i-t)
            SL  2   1/8,1/4,0     Pmmm (1a-h)
            SL  3   1/4,1/8,0     Pmmm (1a-h)
16 n m      SL  1   1/4,0,z       Pmmm (2i-t)   x,0,1/4       Pmmm (2i-t)
            SL  2   1/8,0,1/4     Pmmm (1a-h)
            SL  4   1/4,0,1/8     Pmmm (1a-h)
16 m m      SL  1   0,y,1/4       Pmmm (2i-t)   0,1/4,z       Pmmm (2i-t)
            SL  3   0,1/8,1/4     Pmmm (1a-h)
            SL  4   0,1/4,1/8     Pmmm (1a-h)
16 l 2      SL  1   x,1/4,1/4     Pmmm (2i-t)
            SL  2   1/8,1/4,1/4   Pmmm (1a-h)
16 k 2      SL  1   1/4,y,1/4     Pmmm (2i-t)
            SL  3   1/4,1/8,1/4   Pmmm (1a-h)
16 j 2      SL  1   1/4,1/4,z     Pmmm (2i-t)
            SL  4   1/4,1/4,1/8   Pmmm (1a-h)
 8 i mm2    SL  1   0,0,1/4       Pmmm (1a-h)
 8 h mm2    SL  1   0,1/4,0       Pmmm (1a-h)
 8 g mm2    SL  1   1/4,0,0       Pmmm (1a-h)
 8 f 222    SL  1   1/4,1/4,1/4   Pmmm (1a-h)
 8 e 2/m    SL  1   1/4,1/4,0     Pmmm (1a-h)
 8 d 2/m    SL  1   1/4,0,1/4     Pmmm (1a-h)
 8 c 2/m    SL  1   0,1/4,1/4     Pmmm (1a-h)
```

Space group   No. 70   Fddd

Superlattices

```
SL  8   I(1/2,1/2,1/2)   density  4   add.gen.   1/4,1/4,1/4
SL 12   A(1/4,1/2,1/2)   density  8   add.gen.   1/4,0,0   0,1/4,1/4
SL 19   B(1/2,1/4,1/2)   density  8   add.gen.   0,1/4,0   1/4,0,1/4
SL 25   C(1/2,1/2,1/4)   density  8   add.gen.   0,0,1/4   1/4,1/4,0
SL 27   F(1/2,1/2,1/2)   density  8   add.gen.   1/2,0,0   0,1/4,1/4   1/4,0,1/4
```

Wyckoff letter   non-characteristic crystallographic orbits

```
  32 h 1    SL 8   3/8,y,1/8     Imm (4e-j)     x,3/8,1/8      Imm (4e-j)
                   1/8,3/8,z     Imm (4e-j)
            SL12   0,1/8,3/8     Ammm (2a-d)
            SL19   3/8,0,1/8     Bmmm (2a-d)
            SL25   3/8,1/8,0     Cmmm (2a-d)
            SL27   0,0,1/4       Fmmm (4a,b)
  16 g 2    SL 8   1/8,1/8,3/8   Imm (2a-d)
  16 f 2    SL 8   1/8,3/8,1/8   Imm (2a-d)
  16 e 2    SL 8   3/8,1/8,1/8   Imm (2a-d)
```

Space group No. 71 Immm

Superlattices

```
SL  1  A(1/2,1,1)      density  2  add.gen.  1/2,0,0
SL  2  B(1,1/2,1)      density  2  add.gen.  0,1/2,0
SL  3  C(1,1,1/2)      density  2  add.gen.  0,0,1/2
SL  4  P(1/2,1/2,1/2)  density  4  add.gen.  1/2,0,0   0,1/2,0
SL 11  P(1/4,1/2,1/2)  density  8  add.gen.  1/4,0,0   0,1/2,0
SL 12  P(1/2,1/4,1/2)  density  8  add.gen.  1/2,0,0   0,1/4,0
SL 13  P(1/2,1/2,1/4)  density  8  add.gen.  1/2,0,0   0,0,1/4
```

Wyckoff letter   non-characteristic crystallographic orbits

```
16 o 1     SL 1   1/4,y,z       Ammm (8p,q)
           SL 2   x,1/4,z       Bmmm (8p,q)
           SL 3   x,y,1/4       Cmmm (8p,q)
           SL 4   1/4,1/4,z     Pmmm (2i-t)      1/4,y,1/4    Pmmm (2i-t)
                  x,1/4,1/4     Pmmm (2i-t)
           SL11   1/8,1/4,1/4   Pmmm (1a-h)
           SL12   1/4,1/8,1/4   Pmmm (1a-h)
           SL13   1/4,1/4,1/8   Pmmm (1a-h)
 8 n m     SL 1   1/4,y,0       Ammm (4g-j)
           SL 2   x,1/4,0       Bmmm (4g-j)
           SL 4   1/4,1/4,0     Pmmm (1a-h)
 8 m m     SL 1   1/4,0,z       Ammm (4g-j)
           SL 3   x,0,1/4       Cmmm (4g-j)
           SL 4   1/4,0,1/4     Pmmm (1a-h)
 8 l m     SL 2   0,1/4,z       Bmmm (4g-j)
           SL 3   0,y,1/4       Cmmm (4g-j)
           SL 4   0,1/4,1/4     Pmmm (1a-h)
 8 k 1̄     SL 4   1/4,1/4,1/4   Pmmm (1a-h)
 4 j mm2   SL 3   1/2,0,1/4     Cmmm (2a-d)
 4 i mm2   SL 3   0,0,1/4       Cmmm (2a-d)
 4 h mm2   SL 2   0,1/4,1/2     Bmmm (2a-d)
 4 g mm2   SL 2   0,1/4,0       Bmmm (2a-d)
 4 f mm2   SL 1   1/4,1/2,0     Ammm (2a-d)
 4 e mm2   SL 1   1/4,0,0       Ammm (2a-d)
```

Space group   No. 72   Ibam

Superlattices

SL				density		add.gen.		
SL	1	A(1/2,1,1)	density	2	add.gen.	1/2,0,0	
SL	2	B(1,1/2,1)	density	2	add.gen.	0,1/2,0	
SL	3	C(1,1,1/2)	density	2	add.gen.	0,0,1/2	
SL	4	P(1/2,1/2,1/2)	density	4	add.gen.	1/2,0,0	0,1/2,0
SL	6	F(1/2,1,1)	density	4	add.gen.	1/4,0,1/2	
SL	8	F(1,1/2,1)	density	4	add.gen.	0,1/4,1/2	
SL	9	C(1,1,1/4)	density	4	add.gen.	0,0,1/4	
SL	11	P(1/4,1/2,1/2)	density	8	add.gen.	1/4,0,0	0,1/2,0
SL	12	P(1/2,1/4,1/2)	density	8	add.gen.	1/2,0,0	0,1/4,0
SL	13	P(1/2,1/2,1/4)	density	8	add.gen.	1/2,0,0	0,0,1/4

Wyckoff letter   non-characteristic crystallographic orbits

16 k 1	SL 1	x,1/4,z	Abmm (8m,n)			
	SL 2	1/4,y,z	Bmam (8m,n)			
	SL 3	x,y,1/4	Cmmm (8p,q)	x,0,z	Cmmm (8n,o)	
		0,y,z	Cmmm (8n,o)			
	SL 4	1/4,1/4,z	Pmmm (2i-t)	x,1/4,1/4	Pmmm (2i-t)	
		0,1/4,z	Pmmm (2i-t)	1/4,y,1/4	Pmmm (2i-t)	
		1/4,0,z	Pmmm (2i-t)			
	SL 6	1/8,1/4,z	Fmmm (8g-i)			
	SL 8	1/4,1/8,z	Fmmm (8g-i)			
	SL 9	0,y,1/8	Cmmm (4g-j)	x,0,1/8	Cmmm (4g-j)	
	SL11	1/8,1/4,1/4	Pmmm (1a-h)			
	SL12	1/4,1/8,1/4	Pmmm (1a-h)			
	SL13	1/4,1/4,1/8	Pmmm (1a-h)	0,1/4,1/8	Pmmm (1a-h)	
		1/4,0,1/8	Pmmm (1a-h)			
8 j m	SL 1	x,1/4,0	Abmm (4g)			
	SL 2	1/4,y,0	Bmam (4g)			
	SL 3	x,0,0	Cmmm (4g-j)	0,y,0	Cmmm (4g-j)	
	SL 4	1/4,1/4,0	Pmmm (1a-h)	0,1/4,0	Pmmm (1a-h)	
		1/4,0,0	Pmmm (1a-h)			
	SL 6	1/8,1/4,0	Fmmm (4a,b)	3/8,1/4,0	Fmmm (4a,b)	
	SL 8	1/4,1/8,0	Fmmm (4a,b)	1/4,3/8,0	Fmmm (4a,b)	
8 i 2	SL 3	0,1/2,z	Cmmm (4k,l)			
	SL 9	0,1/2,1/8	Cmmm (2a-d)			
8 h 2	SL 3	0,0,z	Cmmm (4k,l)			
	SL 9	0,0,1/8	Cmmm (2a-d)			
8 g 2	SL 3	0,y,1/4	Cmmm (4g-j)			
	SL 4	0,1/4,1/4	Pmmm (1a-h)			
8 f 2	SL 3	x,0,1/4	Cmmm (4g-j)			
	SL 4	1/4,0,1/4	Pmmm (1a-h)			
8 e $\overline{1}$	SL 4	1/4,1/4,1/4	Pmmm (1a-h)			
4 d 2/m	SL 3	1/2,0,0	Cmmm (2a-d)			
4 c 2/m	SL 3	0,0,0	Cmmm (2a-d)			
4 b 222	SL 3	1/2,0,1/4	Cmmm (2a-d)			
4 a 222	SL 3	0,0,1/4	Cmmm (2a-d)			

Space group   No. 73   Ibca

Superlattices

```
SL  1   A(1/2,1,1)       density  2  add.gen.  1/2,0,0
SL  2   B(1,1/2,1)       density  2  add.gen.  0,1/2,0
SL  3   C(1,1,1/2)       density  2  add.gen.  0,0,1/2
SL  4   P(1/2,1/2,1/2)   density  4  add.gen.  1/2,0,0   0,1/2,0
SL  6   F(1/2,1,1)       density  4  add.gen.  1/4,0,1/2
SL  8   F(1,1/2,1)       density  4  add.gen.  0,1/4,1/2
SL 10   F(1,1,1/2)       density  4  add.gen.  0,1/2,1/4
SL 11   P(1/4,1/2,1/2)   density  8  add.gen.  1/4,0,0   0,1/2,0
SL 12   P(1/2,1/4,1/2)   density  8  add.gen.  1/2,0,0   0,1/4,0
SL 13   P(1/2,1/2,1/4)   density  8  add.gen.  1/2,0,0   0,0,1/4
```

Wyckoff letter   non-characteristic crystallographic orbits

```
 16 f 1    SL  1   x,y,1/4        Abmm (8m,n)     x,0,z         Abmm (8m,n)
           SL  2   1/4,y,z        Bmcm (8m,n)     x,y,0         Bmcm (8m,n)
           SL  3   x,1/4,z        Cmma (8m,n)     0,y,z         Cmma (8m,n)
           SL  4   x,1/4,1/4      Pmmm (2i-t)     0,y,1/4       Pmmm (2i-t)
                   1/4,y,1/4      Pmmm (2i-t)     1/4,1/4,z     Pmmm (2i-t)
                   1/4,0,z        Pmmm (2i-t)     x,0,0         Pmmm (2i-t)
                   0,0,z          Pmmm (2i-t)     x,1/4,0       Pmmm (2i-t)
                   0,y,0          Pmmm (2i-t)
           SL  6   1/8,y,1/4      Fmmm (8g-i)     1/8,0,z       Fmmm (8g-i)
           SL  8   x,1/8,0        Fmmm (8g-i)     1/4,1/8,z     Fmmm (8g-i)
           SL 10   x,1/4,1/8      Fmmm (8g-i)     0,y,1/8       Fmmm (8g-i)
           SL 11   1/8,1/4,1/4    Pmmm (1a-h)     1/8,0,0       Pmmm (1a-h)
                   1/8,1/4,0      Pmmm (1a-h)
           SL 12   1/4,1/8,1/4    Pmmm (1a-h)     0,1/8,1/4     Pmmm (1a-h)
                   0,1/8,0        Pmmm (1a-h)
           SL 13   1/4,1/4,1/8    Pmmm (1a-h)     0,0,1/8       Pmmm (1a-h)
                   1/4,0,1/8      Pmmm (1a-h)
  8 e 2    SL  3   0,1/4,z        Cmma (4g)
           SL  4   0,1/4,1/4      Pmmm (1a-h)     0,1/4,0       Pmmm (1a-h)
           SL 10   0,1/4,3/8      Fmmm (4a,b)     0,1/4,1/8     Fmmm (4a,b)
  8 d 2    SL  2   1/4,y,0        Bmcm (4g)
           SL  4   1/4,0,0        Pmmm (1a-h)     1/4,1/4,0     Pmmm (1a-h)
           SL  8   1/4,1/8,0      Fmmm (4a,b)     1/4,3/8,0     Fmmm (4a,b)
  8 c 2    SL  1   x,0,1/4        Abmm (4g)
           SL  4   1/4,0,1/4      Pmmm (1a-h)     0,0,1/4       Pmmm (1a-h)
           SL  6   3/8,0,1/4      Fmmm (4a,b)     1/8,0,1/4     Fmmm (4a,b)
  8 b 1̄   SL  4   1/4,1/4,1/4    Pmmm (1a-h)
  8 a 1̄   SL  4   0,0,0          Pmmm (1a-h)
```

Space group No. 74 Imma

Superlattices

```
SL  1  A(1/2,1,1)      density  2  add.gen.  1/2,0,0
SL  2  B(1,1/2,1)      density  2  add.gen.  0,1/2,0
SL  4  P(1/2,1/2,1/2)  density  4  add.gen.  1/2,0,0   0,1/2,0
SL  5  A(1/4,1,1)      density  4  add.gen.  1/4,0,0
SL  7  B(1,1/4,1)      density  4  add.gen.  0,1/4,0
SL 11  P(1/4,1/2,1/2)  density  8  add.gen.  1/4,0,0   0,1/2,0
SL 12  P(1/2,1/4,1/2)  density  8  add.gen.  1/2,0,0   0,1/4,0
SL 13  P(1/2,1/2,1/4)  density  8  add.gen.  1/2,0,0   0,0,1/4
```

Wyckoff letter   non-characteristic crystallographic orbits

```
 16 j 1      SL  1   x,y,1/4        Ammm (8n,o)    1/4,y,z        Ammm (8p,q)
             SL  2   x,0,z          Bmmm (8p,q)    x,y,0          Bmmm (8n,o)
             SL  4   x,0,1/4        Pmmm (2i-t)    1/4,0,z        Pmmm (2i-t)
                     1/4,y,0        Pmmm (2i-t)
             SL  5   1/8,y,1/4      Ammm (4g-j)
             SL  7   x,1/8,0        Bmmm (4g-j)
             SL11    1/8,0,1/4      Pmmm (1a-h)
             SL12    1/4,1/8,0      Pmmm (1a-h)
             SL13    1/4,0,1/8      Pmmm (1a-h)
  8 i m      SL  1   x,1/4,3/4      Ammm (4k,l)    x,1/4,1/4      Ammm (4k,l)
                     1/4,1/4,z      Ammm (4g-j)
             SL  2   x,1/4,0        Bmmm (4g-j)
             SL  4   1/4,1/4,0      Pmmm (1a-h)
             SL  5   1/8,1/4,3/4    Ammm (2a-d)    1/8,1/4,1/4    Ammm (2a-d)
  8 h m      SL  1   0,y,1/4        Ammm (4g-j)
             SL  2   0,0,z          Bmmm (4g-j)    0,y,1/2        Bmmm (4k,l)
                     0,y,0          Bmmm (4k,l)
             SL  4   0,0,1/4        Pmmm (1a-h)
             SL  7   0,1/8,1/2      Bmmm (2a-d)    0,1/8,0        Bmmm (2a-d)
  8 g 2      SL  1   1/4,y,1/4      Ammm (4g-j)
             SL  4   1/4,0,1/4      Pmmm (1a-h)
  8 f 2      SL  2   x,0,0          Bmmm (4g-j)
             SL  4   1/4,0,0        Pmmm (1a-h)
  4 e mm2    SL  1   0,1/4,3/4      Ammm (2a-d)    0,1/4,1/4      Ammm (2a-d)
             SL  2   0,1/4,1/2      Bmmm (2a-d)    0,1/4,0        Bmmm (2a-d)
  4 d 2/m    SL  1   1/4,1/4,3/4    Ammm (2a-d)
  4 c 2/m    SL  1   1/4,1/4,1/4    Ammm (2a-d)
  4 b 2/m    SL  2   0,0,1/2        Bmmm (2a-d)
  4 a 2/m    SL  2   0,0,0          Bmmm (2a-d)
```

3.4. Tetragonal system

Space group No. 75 P4

Superlattices

```
SL  0  P(1,1,1)          density  1  add.gen.
SL  2  C(1,1,1)          density  2  add.gen.   1/2,1/2,0
SL  7  P(1/2,1/2,1)      density  4  add.gen.   1/2,0,0  0,1/2,0
```

Wyckoff letter   non-characteristic crystallographic orbits

```
   4 d 1     SL 0   x,y,z        P4/m (4j,k)      x,0,z      P4/mmm (4l-o)
                    x,1/2,z      P4/mmm (4l-o)    x,x,z      P4/mmm (4j,k)
                    x,1/2+x,z    P4/mbm (4g,h)
             SL 7   1/4,1/4,z    P4/mmm (1a-d)
   2 c 2     SL 2   0,1/2,z      P4/mmm (1a-d)
   1 b 4     SL 0   1/2,1/2,z    P4/mmm (1a-d)
   1 a 4     SL 0   0,0,z        P4/mmm (1a-d)
```

Space group No. 76 P4₁

Superlattices

```
SL  0  P(1,1,1)          density  1  add.gen.
SL  3  I(1,1,1)          density  2  add.gen.   1/2,1/2,1/2
SL  4  P(1,1,1/4)        density  4  add.gen.   0,0,1/4
SL  6  I(1,1,1/2)        density  4  add.gen.   1/2,1/2,1/4
```

Wyckoff letter   non-characteristic crystallographic orbits

```
   4 a 1     SL 0   x,x,z        P4₁22 (4c)       x,1/2,z     P4₁22 (4a,b)
                    x,0,z        P4₁22 (4a,b)     x,1/2+x,z   P4₁2₁2 (4a)
             SL 3   1/4,1/4,z    I4₁/amd (4a,b)
             SL 4   1/2,1/2,z    P4/mmm (1a-d)    0,0,z       P4/mmm (1a-d)
             SL 6   1/2,0,z      I4/mmm (2a,b)
```

Space group No. 77 P4₂

Superlattices

```
SL  0  P(1,1,1)        density  1  add.gen.
SL  1  P(1,1,1/2)      density  2  add.gen.  0,0,1/2
SL  3  I(1,1,1)        density  2  add.gen.  1/2,1/2,1/2
SL  8  F(1,1,1)        density  4  add.gen.  0,1/2,1/2  1/2,0,1/2
```

Wyckoff letter   non-characteristic crystallographic orbits

```
   4 d 1    SL 0   x,y,z        P4₂/m (4j)      x,x,z       P4₂/mcm (4i,j)
                   x,1/2+x,z    P4₂/mnm (4f,g)  x,0,z       P4₂/mmc (4j-m)
                   x,1/2,z      P4₂/mmc (4j-m)
            SL 8   1/4,1/4,z    I4/mmm (2a,b)
   2 c 2    SL 3   0,1/2,z      I4/mmm (2a,b)
   2 b 2    SL 1   1/2,1/2,z    P4/mmm (1a-d)
   2 a 2    SL 1   0,0,z        P4/mmm (1a-d)
```

Space group No. 78 P4₃

Superlattices

```
SL  0  P(1,1,1)        density  1  add.gen.
SL  3  I(1,1,1)        density  2  add.gen.  1/2,1/2,1/2
SL  4  P(1,1,1/4)      density  4  add.gen.  0,0,1/4
SL  6  I(1,1,1/2)      density  4  add.gen.  1/2,1/2,1/4
```

Wyckoff letter   non-characteristic crystallographic orbits

```
   4 a 1    SL 0   x,x,z        P4₃22 (4c)      x,1/2,z       P4₃22 (4a,b)
                   x,0,z        P4₃22 (4a,b)    x,1/2+x,z     P4₃2₁2 (4a)
            SL 3   1/4,1/4,z    I4₁/amd (4a,b)
            SL 4   1/2,1/2,z    P4/mmm (1a-d)   0,0,z         P4/mmm (1a-d)
            SL 6   1/2,0,z      I4/mmm (2a,b)
```

Space group   No. 79   I4

Superlattices

```
SL  0  I(1,1,1)        density  1  add.gen.
SL  1  C(1,1,1/2)      density  2  add.gen.  0,0,1/2
SL  3  P(1/2,1/2,1/2)  density  4  add.gen.  1/2,0,0  0,1/2,0
```

Wyckoff letter   non-characteristic crystallographic orbits

```
 8 c 1     SL 0   x,y,z        I4/m (8h)        x,0,z      I4/mmm (8i,j)
                  x,1/2,z      I4/mmm (8i,j)    x,x,z      I4/mmm (8h)
                  x,1/2+x,z    I4/mcm (8h)
           SL 3   1/4,1/4,z    P4/mmm (1a-d)
 4 b 2     SL 1   0,1/2,z      P4/mmm (1a-d)
 2 a 4     SL 0   0,0,z        I4/mmm (2a,b)
```

Space group   No. 80   I4$_1$

Superlattices

```
SL  0  I(1,1,1)        density  1  add.gen.
SL  2  C(1,1,1/4)      density  4  add.gen.  0,0,1/4
```

Wyckoff letter   non-characteristic crystallographic orbits

```
 8 b 1     SL 0   x,1/4,z      I4$_1$22 (8f)     x,-x,z      I4$_1$22 (8d,e)
                  x,x,z        I4$_1$22 (8d,e)   x,0,z       I4$_1$md (8b)
                  0,1/4,z      I4$_1$/amd (8c,d)
           SL 2   3/4,1/4,z    P4/mmm (1a-d)    1/4,1/4,z   P4/mmm (1a-d)
 4 a 2     SL 0   0,0,z        I4$_1$/amd (4a,b)
```

Space group No. 81 P$\bar{4}$

Superlattices

SL	0	P(1,1,1)	density	1	add.gen.		
SL	1	P(1,1,1/2)	density	2	add.gen.	0,0,1/2	
SL	2	C(1,1,1)	density	2	add.gen.	1/2,1/2,0	
SL	3	I(1,1,1)	density	2	add.gen.	1/2,1/2,1/2	
SL	7	P(1/2,1/2,1)	density	4	add.gen.	1/2,0,0 0,1/2,0	
SL	8	F(1,1,1)	density	4	add.gen.	0,1/2,1/2 1/2,0,1/2	

Wyckoff letter   non-characteristic crystallographic orbits

Wyckoff	SL	coords	group	coords	group
4 h 1	SL 0	x,y,0	P4/m (4j,k)	x,y,1/2	P4/m (4j,k)
		x,y,1/4	P4$_2$/m (4j)	x,x,z	P$\bar{4}$2m (4n)
		x,1/2+x,z	P$\bar{4}$2$_1$m (4e)	x,1/2,z	P$\bar{4}$m2 (4j,k)
		x,0,z	P$\bar{4}$m2 (4j,h)	x,1/2,0	P4/mmm (4l-o)
		x,0,0	P4/mmm (4l-o)	x,1/2,1/2	P4/mmm (4l-o)
		x,0,1/2	P4/mmm (4l-o)	x,x,0	P4/mmm (4j,k)
		x,x,1/2	P4/mmm (4j,k)	x,0,3/4	P4$_2$/mmc (4j-m)
		x,1/2,1/4	P4$_2$/mmc (4j-m)	x,1/2,3/4	P4$_2$/mmc (4j-m)
		x,0,1/4	P4$_2$/mmc (4j-m)	x,x,1/4	P4$_2$/mcm (4i,j)
		x,x,3/4	P4$_2$/mcm (4i,j)	x,1/2+x,1/2	P4/mbm (4g,h)
		x,1/2+x,0	P4/mbm (4g,h)	x,1/2+x,1/4	P4$_2$/mnm (4f,g)
		x,1/2+x,3/4	P4$_2$/mnm (4f,g)		
	SL 2	1/4,1/4,z	P4/nmm (2c)		
	SL 7	1/4,1/4,1/2	P4/mmm (1a-d)	1/4,1/4,0	P4/mmm (1a-d)
	SL 8	1/4,1/4,3/4	I4/mmm (2a,b)	1/4,1/4,1/4	I4/mmm (2a,b)
2 g 2	SL 0	0,1/2,z	P4/nmm (2c)		
	SL 2	0,1/2,1/2	P4/mmm (1a-d)	0,1/2,0	P4/mmm (1a-d)
	SL 3	0,1/2,1/4	I4/mmm (2a,b)	0,1/2,3/4	I4/mmm (2a,b)
2 f 2	SL 0	1/2,1/2,z	P4/mmm (2g,h)		
	SL 1	1/2,1/2,1/4	P4/mmm (1a-d)		
2 e 2	SL 0	0,0,z	P4/mmm (2g,h)		
	SL 1	0,0,1/4	P4/mmm (1a-d)		
1 d $\bar{4}$	SL 0	1/2,1/2,1/2	P4/mmm (1a-d)		
1 c $\bar{4}$	SL 0	1/2,1/2,0	P4/mmm (1a-d)		
1 b $\bar{4}$	SL 0	0,0,1/2	P4/mmm (1a-d)		
1 a $\bar{4}$	SL 0	0,0,0	P4/mmm (1a-d)		

Space group No. 82 I$\bar{4}$

Superlattices

```
SL  0  I(1,1,1)          density  1  add.gen.
SL  1  C(1,1,1/2)        density  2  add.gen.   0,0,1/2
SL  3  P(1/2,1/2,1/2)    density  4  add.gen.   1/2,0,0  0,1/2,0
SL  4  F(1,1,1/2)        density  4  add.gen.   0,1/2,1/4
```

Wyckoff letter   non-characteristic crystallographic orbits

```
8 g 1    SL 0   x,y,0        I4/m (8h)       x,y,1/4      I4/m (8h)
                x,0,z        I4m2 (8i)       x,x,z        I42m (8i)
                x,1/2+x,z    I42m (8i)       x,1/4,1/8    I42d (8d)
                x,1/4,3/8    I42d (8d)       x,x,0        I4/mmm (8h)
                x,0,0        I4/mmm (8i,j)   x,1/2,0      I4/mmm (8i,j)
                x,0,1/4      I4/mmm (8i,j)   x,1/2,1/4    I4/mmm (8i,j)
                x,1/2+x,1/4  I4/mmm (8h)     x,x,1/4      I4/mcm (8h)
                x,1/2+x,0    I4/mcm (8h)     0,1/4,1/8    I4₁/amd (8c,d)
                1/2,1/4,1/8  I4₁/amd (8c,d)  0,1/4,3/8    I4₁/amd (8c,d)
                1/2,1/4,3/8  I4₁/amd (8c,d)
         SL 1   1/4,1/4,z    P4/nmm (2c)
         SL 3   1/4,1/4,0    P4/mmm (1a-d)   1/4,1/4,1/4  P4/mmm (1a-d)
         SL 4   1/4,1/4,3/8  I4/mmm (2a,b)   1/4,1/4,1/8  I4/mmm (2a,b)
  4 f 2  SL 0   0,1/2,z      I4/mmm (4e)
         SL 1   0,1/2,0      P4/mmm (1a-d)
  4 e 2  SL 0   0,0,z        I4/mmm (4e)
         SL 1   0,0,1/4      P4/mmm (1a-d)
  2 d 4̄  SL 0   0,1/2,3/4    I4/mmm (2a,b)
  2 c 4̄  SL 0   0,1/2,1/4    I4/mmm (2a,b)
  2 b 4̄  SL 0   0,0,1/2      I4/mmm (2a,b)
  2 a 4̄  SL 0   0,0,0        I4/mmm (2a,b)
```

Space group No. 83 P4/m

Superlattices

SL	0	P(1,1,1)	density	1	add.gen.			
SL	1	P(1,1,1/2)	density	2	add.gen.	0,0,1/2		
SL	2	C(1,1,1)	density	2	add.gen.	1/2,1/2,0		
SL	5	C(1,1,1/2)	density	4	add.gen.	0,0,1/2	1/2,1/2,0	
SL	7	P(1/2,1/2,1)	density	4	add.gen.	1/2,0,0	0,1/2,0	
SL	12	P(1/2,1/2,1/2)	density	8	add.gen.	1/2,0,0	0,1/2,0	0,0,1/2

Wyckoff letter   non-characteristic crystallographic orbits

Wyckoff	SL	orbit	group	orbit	group
8 l 1	SL 0	x,1/2,z	P4/mmm (8s,t)	x,0,z	P4/mmm (8s,t)
		x,x,z	P4/mmm (8r)	x,1/2+x,z	P4/mbm (8k)
	SL 1	x,y,1/4	P4/m (4j,k)	x,1/2,1/4	P4/mmm (4l-o)
		x,0,1/4	P4/mmm (4l-o)	x,x,1/4	P4/mmm (4j,k)
		x,1/2+x,1/4	P4/mbm (4g,h)		
	SL 7	1/4,1/4,z	P4/mmm (2g,h)		
	SL 12	1/4,1/4,1/4	P4/mmm (1a-d)		
4 k m	SL 0	x,1/2,1/2	P4/mmm (4l-o)	x,0,1/2	P4/mmm (4l-o)
		x,x,1/2	P4/mmm (4j,k)	x,1/2+x,1/2	P4/mbm (4g,h)
	SL 7	1/4,1/4,1/2	P4/mmm (1a-d)		
4 j m	SL 0	x,1/2,0	P4/mmm (4l-o)	x,0,0	P4/mmm (4l-o)
		x,x,0	P4/mmm (4j,k)	x,1/2+x,0	P4/mbm (4g,h)
	SL 7	1/4,1/4,0	P4/mmm (1a-d)		
4 i 2	SL 2	0,1/2,z	P4/mmm (2g,h)		
	SL 5	0,1/2,1/4	P4/mmm (1a-d)		
2 h 4	SL 0	1/2,1/2,z	P4/mmm (2g,h)		
	SL 1	1/2,1/2,1/4	P4/mmm (1a-d)		
2 g 4	SL 0	0,0,z	P4/mmm (2g,h)		
	SL 1	0,0,1/4	P4/mmm (1a-d)		
2 f 2/m	SL 2	0,1/2,1/2	P4/mmm (1a-d)		
2 e 2/m	SL 2	0,1/2,0	P4/mmm (1a-d)		
1 d 4/m	SL 0	1/2,1/2,1/2	P4/mmm (1a-d)		
1 c 4/m	SL 0	1/2,1/2,0	P4/mmm (1a-d)		
1 b 4/m	SL 0	0,0,1/2	P4/mmm (1a-d)		
1 a 4/m	SL 0	0,0,0	P4/mmm (1a-d)		

Space group No. 84   P4₂/m

Superlattices

SL	0	P(1,1,1)	density	1	add.gen.	
SL	1	P(1,1,1/2)	density	2	add.gen.	0,0,1/2
SL	3	I(1,1,1)	density	2	add.gen.	1/2,1/2,1/2
SL	4	P(1,1,1/4)	density	4	add.gen.	0,0,1/4
SL	5	C(1,1,1/2)	density	4	add.gen.	0,0,1/2 1/2,1/2,0
SL	8	F(1,1,1)	density	4	add.gen.	0,1/2,1/2 1/2,0,1/2
SL	12	P(1/2,1/2,1/2)	density	8	add.gen.	1/2,0,0 0,1/2,0 0,0,1/2

Wyckoff letter   non-characteristic crystallographic orbits

Wyckoff		SL	coord	group	coord	group
8 k	1	SL 0	x,0,z	P4₂/mmc (8o,p)	x,1/2,z	P4₂/mmc (8o,p)
			x,x,z	P4₂/mcm (8o)	x,1/2+x,z	P4₂/mnm (8j)
		SL 1	x,y,1/4	P4/m (4j,k)	x,x,1/4	P4/mmm (4j,k)
			x,1/2,1/4	P4/mmm (4l-o)	x,0,1/4	P4/mmm (4l-o)
			x,1/2+x,1/4	P4/mbm (4g,h)		
		SL 8	1/4,1/4,z	I4/mmm (4e)		
		SL12	1/4,1/4,1/4	P4/mmm (1a-d)		
4 j	m	SL 0	x,1/2,0	P4₂/mmc (4j-m)	1/2,y,0	P4₂/mmc (4j-m)
			x,0,0	P4₂/mmc (4j-m)	0,y,0	P4₂/mmc (4j-m)
			x,-x,0	P4₂/mcm (i,j)	x,x,0	P4₂/mcm (4i,j)
			x,1/2-x,0	P4₂/mnm (4f,g)	x,1/2+x,0	P4₂/mnm (4f,g)
		SL 8	1/4,3/4,0	I4/mmm (2a,b)	1/4,1/4,0	I4/mmm (2a,b)
4 i	2	SL 3	0,1/2,z	I4/mmm (4e)		
		SL 5	0,1/2,1/4	P4/mmm (1a-d)		
4 h	2	SL 1	1/2,1/2,z	P4/mmm (2g,h)		
		SL 4	1/2,1/2,1/8	P4/mmm (1a-d)		
4 g	2	SL 1	0,0,z	P4/mmm (2g,h)		
		SL 4	0,0,1/8	P4/mmm (1a-d)		
2 f	4̄	SL 1	1/2,1/2,1/4	P4/mmm (1a-d)		
2 e	4̄	SL 1	0,0,1/4	P4/mmm (1a-d)		
2 d	2/m	SL 3	0,1/2,1/2	I4/mmm (2a,b)		
2 c	2/m	SL 3	0,1/2,0	I4/mmm (2a,b)		
2 b	2/m	SL 1	1/2,1/2,0	P4/mmm (1a-d)		
2 a	2/m	SL 1	0,0,0	P4/mmm (1a-d)		

Space group No. 85   P4/n

Superlattices

```
SL  0   P(1,1,1)        density  1  add.gen.
SL  2   C(1,1,1)        density  2  add.gen.  1/2,1/2,0
SL  3   I(1,1,1)        density  2  add.gen.  1/2,1/2,1/2
SL  5   C(1,1,1/2)      density  4  add.gen.  0,0,1/2    1/2,1/2,0
SL  7   P(1/2,1/2,1)    density  4  add.gen.  1/2,0,0    0,1/2,0
SL 12   P(1/2,1/2,1/2)  density  8  add.gen.  1/2,0,0    0,1/2,0    0,0,1/2
SL 14   C(1/2,1/2,1)    density  8  add.gen.  1/2,0,0    1/4,1/4,0
```

Wyckoff letter   non-characteristic crystallographic orbits

8 g 1	SL 0	x,-x,z	P4/nbm (8m)	x,x,z	P4/nmm (8j)
		x,1/4,z	P4/nmm (8i)		
	SL 2	x,y,1/2	P4/m (4j,k)	x,y,0	P4/m (4j,k)
		x,1/4,0	P4/mmm (4j,k)	x,1/4,1/2	P4/mmm (4j,k)
		x,x,0	P4/mmm (4l-o)	x,x,1/2	P4/mmm (4l-o)
		x,1/2+x,0	P4/mmm (4l-o)	x,1/2+x,1/2	P4/mmm (4l-o)
		x,0,0	P4/mbm (4g,h)	x,0,1/2	P4/mbm (4g,h)
	SL 3	x,y,1/4	I4/m (8h)	x,x,1/4	I4/mmm (8h)
		x,1/2+x,1/4	I4/mcm (8h)	x,1/4,1/4	I4/mmm (8i,j)
		x,3/4,1/4	I4/mmm (8i,j)		
	SL 7	0,0,z	P4/mmm (2g,h)		
	SL12	0,0,1/4	P4/mmm (1a-d)		
	SL14	1/4,0,1/2	P4/mmm (1a-d)	1/4,0,0	P4/mmm (1a-d)
4 f 2	SL 2	1/4,3/4,z	P4/mmm (2g,h)		
	SL 5	1/4,3/4,1/4	P4/mmm (1a-d)		
4 e $\bar{1}$	SL 7	0,0,1/2	P4/mmm (1a-d)		
4 d $\bar{1}$	SL 7	0,0,0	P4/mmm (1a-d)		
2 c 4	SL 0	1/4,1/4,z	P4/nmm (2c)		
	SL 2	1/4,1/4,1/2	P4/mmm (1a-d)	1/4,1/4,0	P4/mmm (1a-d)
	SL 3	1/4,1/4,1/4	I4/mmm (2a,b)	1/4,1/4,3/4	I4/mmm (2a,b)
2 b $\bar{4}$	SL 2	1/4,3/4,1/2	P4/mmm (1a-d)		
2 a $\bar{4}$	SL 2	1/4,3/4,0	P4/mmm (1a-d)		

Space group No. 86 P4$_2$/n

Superlattices

```
SL  0  P(1,1,1)          density 1  add.gen.
SL  1  P(1,1,1/2)        density 2  add.gen.  0,0,1/2
SL  2  C(1,1,1)          density 2  add.gen.  1/2,1/2,0
SL  3  I(1,1,1)          density 2  add.gen.  1/2,1/2,1/2
SL  5  C(1,1,1/2)        density 4  add.gen.  0,0,1/2   1/2,1/2,0
SL  6  I(1,1,1/2)        density 4  add.gen.  1/2,1/2,1/4
SL  8  F(1,1,1)          density 4  add.gen.  0,1/2,1/2  1/2,0,1/2
SL 12  P(1/2,1/2,1/2)    density 8  add.gen.  1/2,0,0   0,1/2,0   0,0,1/2
SL 15  I(1/2,1/2,1)      density 8  add.gen.  1/2,0,0   1/4,1/4,1/2
```

Wyckoff letter   non-characteristic crystallographic orbits

8 g 1	SL 0	x,x,z	P4$_2$/nnm (8m)	x,1/4,z	P4$_2$/nmc (8g)
		x,-x,z	P4$_2$/ncm (8i)		
	SL 2	x,y,0	P4$_2$/m (4j)	x,-x,0	P4$_2$/mmc (4j-m)
		x,-x,1/2	P4$_2$/mmc (4j-m)	x,1/2-x,1/2	P4$_2$/mmc (4j-m)
		x,1/2-x,0	P4$_2$/mmc (4j-m)	x,1/4,1/2	P4$_2$/mcm (4i,j)
		x,1/4,0	P4$_2$/mcm (4i,j)	x,0,0	P4$_2$/mnm (4f,g)
		x,0,1/2	P4$_2$/mnm (4f,g)		
	SL 3	x,y,1/4	I4/m (8h)	x,1/4,1/4	I4/mmm (8i,j)
		x,1/4,3/4	I4/mmm (8i,j)	x,x,1/4	I4/mmm (8h)
		x,-x,1/4	I4/mcm (8h)		
	SL 8	0,0,z	I4/mmm (4e)		
	SL12	0,0,1/4	P4/mmm (1a-d)		
	SL15	0,1/4,0	I4/mmm (2a,b)	1/4,0,0	I4/mmm (2a,b)
4 f 2	SL 3	1/4,1/4,z	I4/mmm (4e)		
	SL 5	1/4,1/4,0	P4/mmm (1a-d)		
4 e 2	SL 1	3/4,1/4,z	P4/nmm (2c)		
	SL 5	3/4,1/4,1/4	P4/mmm (1a-d)	3/4,1/4,0	P4/mmm (1a-d)
	SL 6	3/4,1/4,3/8	I4/mmm (2a,b)	3/4,1/4,1/8	I4/mmm (2a,b)
4 d $\bar{1}$	SL 8	0,0,1/2	I4/mmm (2a,b)		
4 c $\bar{1}$	SL 8	0,0,0	I4/mmm (2a,b)		
2 b $\bar{4}$	SL 3	1/4,1/4,3/4	I4/mmm (2a,b)		
2 a $\bar{4}$	SL 3	1/4,1/4,1/4	I4/mmm (2a,b)		

Space group No. 87 I4/m

Superlattices

```
SL  0   I(1,1,1)          density  1   add.gen.
SL  1   C(1,1,1/2)        density  2   add.gen.  0,0,1/2
SL  2   C(1,1,1/4)        density  4   add.gen.  0,0,1/4
SL  3   P(1/2,1/2,1/2)    density  4   add.gen.  1/2,0,0   0,1/2,0
SL  6   P(1/2,1/2,1/4)    density  8   add.gen.  1/2,0,0   0,0,1/4
SL  8   C(1/2,1/2,1/2)    density  8   add.gen.  1/2,0,0   1/4,1/4,0
```

Wyckoff letter   non-characteristic crystallographic orbits

```
16 i 1    SL 0   0,y,z          I4/mmm (16n)    x,x,z          I4/mmm (16m)
                 x,1/2+x,z      I4/mcm (16l)
          SL 1   x,y,1/4        P4/m (4j,k)     x,0,1/4        P4/mmm (4j,k)
                 x,x,1/4        P4/mmm (4l-o)   x,1/2+x,1/4    P4/mmm (4l-o)
                 x,1/4,1/4      P4/mbm (4g,h)
          SL 3   1/4,1/4,z      P4/mmm (2g,h)
          SL 6   1/4,1/4,1/8    P4/mmm (1a-d)
          SL 8   1/4,0,1/4      P4/mmm (1a-d)
 8 h m    SL 0   x,1/2,0        I4/mmm (8i,j)   x,0,0          I4/mmm (8i,j)
                 x,x,0          I4/mmm (8h)     x,1/2+x,0      I4/mcm (8h)
          SL 3   1/4,1/4,0      P4/mmm (1a-d)
 8 g 2    SL 1   0,1/2,z        P4/mmm (2g,h)
          SL 2   0,1/2,1/8      P4/mmm (1a-d)
 8 f 1̄    SL 3   1/4,1/4,1/4    P4/mmm (1a-d)
 4 e 4    SL 0   0,0,z          I4/mmm (4e)
          SL 1   0,0,1/4        P4/mmm (1a-d)
 4 d 4̄    SL 1   0,1/2,1/4      P4/mmm (1a-d)
 4 c 2/m  SL 1   0,1/2,0        P4/mmm (1a-d)
 2 b 4/m  SL 0   0,0,1/2        I4/mmm (2a,b)
 2 a 4/m  SL 0   0,0,0          I4/mmm (2a,b)
```

Space group   No. 88   I4$_1$/a

Superlattices

```
SL  0  I(1,1,1)          density  1  add.gen.
SL  2  C(1,1,1/4)        density  4  add.gen.  0,0,1/4
SL  4  F(1,1,1/2)        density  4  add.gen.  0,1/2,1/4
SL  6  P(1/2,1/2,1/4)    density  8  add.gen.  1/2,0,0  0,0,1/4
SL  7  F(1,1,1/4)        density  8  add.gen.  0,1/2,1/8
SL  9  I(1/2,1/2,1/2)    density  8  add.gen.  1/2,0,0 0,1/2,0 1/4,1/4,1/4
```

Wyckoff letter		non-characteristic crystallographic orbits			
16 f 1	SL 0	x,1/4+x,1/8	I4$_1$/amd (16g)	x,1/4,z	I4$_1$/amd (16h)
		x,0,0	I4$_1$/amd (16f)	x,0,1/4	I4$_1$/acd (16e)
		x,1/4+x,3/8	I4$_1$/acd (16f)		
	SL 2	1/4,0,z	P4/nmm (2c)		
	SL 6	1/4,0,1/8	P4/mmm (1a-d)	1/4,0,0	P4/mmm (1a-d)
	SL 7	1/4,0,1/16	I4/mmm (2a,b)	1/4,0,3/16	I4/mmm (2a,b)
	SL 9	0,0,1/4	I4/mmm (2a,b)		
8 e 2	SL 0	0,1/4,z	I4$_1$/amd (8e)		
	SL 4	0,1/4,3/8	I4/mmm (2a,b)		
8 d $\overline{1}$	SL 0	0,0,1/2	I4$_1$/amd (8c,d)		
8 c $\overline{1}$	SL 0	0,0,0	I4$_1$/amd (8c,d)		
4 b $\overline{4}$	SL 0	0,1/4,5/8	I4$_1$/amd (4a,b)		
4 a $\overline{4}$	SL 0	0,1/4,1/8	I4$_1$/amd (4a,b)		

Space group No. 89   P422

Superlattices

```
SL  0  P(1,1,1)       density  1  add.gen.
SL  1  P(1,1,1/2)     density  2  add.gen.  0,0,1/2
SL  2  C(1,1,1)       density  2  add.gen.  1/2,1/2,0
SL  3  I(1,1,1)       density  2  add.gen.  1/2,1/2,1/2
SL  5  C(1,1,1/2)     density  4  add.gen.  0,0,1/2    1/2,1/2,0
SL  7  P(1/2,1/2,1)   density  4  add.gen.  1/2,0,0    0,1/2,0
SL 12  P(1/2,1/2,1/2) density  8  add.gen.  1/2,0,0    0,1/2,0    0,0,1/2
```

Wyckoff letter   non-characteristic crystallographic orbits

```
8 p 1      SL 0   x,1/2,z       P4/mmm (8s,t)   x,0,z        P4/mmm (8s,t)
                  x,x,z         P4/mmm (8r)     x,y,1/2      P4/mmm (8p,q)
                  x,y,0         P4/mmm (8p,q)   x,y,1/4      P4/mcc (8m)
                  x,1/2+x,z     P4/nbm (8m)
           SL 1   x,x,1/4       P4/mmm (4j,k)   x,1/2,1/4    P4/mmm (4l-o)
                  x,0,1/4       P4/mmm (4l-o)
           SL 2   x,1/2+x,1/2   P4/mmm (4l-o)   x,1/2+x,0    P4/mmm (4l-o)
           SL 3   x,1/2+x,1/4   I4/mcm (8h)
           SL 7   1/4,1/4,z     P4/mmm (2g,h)
           SL12   1/4,1/4,1/4   P4/mmm (1a-d)
4 o 2      SL 0   x,1/2,0       P4/mmm (4l-o)
4 n 2      SL 0   x,0,1/2       P4/mmm (4l-o)
4 m 2      SL 0   x,1/2,1/2     P4/mmm (4l-o)
4 l 2      SL 0   x,0,0         P4/mmm (4l-o)
4 k 2      SL 0   x,x,1/2       P4/mmm (4j,k)
           SL 7   1/4,1/4,1/2   P4/mmm (1a-d)
4 j 2      SL 0   x,x,0         P4/mmm (4j,k)
           SL 7   1/4,1/4,0     P4/mmm (1a-d)
4 i 2      SL 2   0,1/2,z       P4/mmm (2g,h)
           SL 5   0,1/2,1/4     P4/mmm (1a-d)
2 h 4      SL 0   1/2,1/2,z     P4/mmm (2g,h)
           SL 1   1/2,1/2,1/4   P4/mmm (1a-d)
2 g 4      SL 0   0,0,z         P4/mmm (2g,h)
           SL 1   0,0,1/4       P4/mmm (1a-d)
2 f 222    SL 2   1/2,0,1/2     P4/mmm (1a-d)
2 e 222    SL 2   1/2,0,0       P4/mmm (1a-d)
1 d 422    SL 0   1/2,1/2,1/2   P4/mmm (1a-d)
1 c 422    SL 0   1/2,1/2,0     P4/mmm (1a-d)
1 b 422    SL 0   0,0,1/2       P4/mmm (1a-d)
1 a 422    SL 0   0,0,0         P4/mmm (1a-d)
```

Space group   No. 90   P4₂₁2

Superlattices

```
SL  0  P(1,1,1)        density  1  add.gen.
SL  1  P(1,1,1/2)      density  2  add.gen.  0,0,1/2
SL  2  C(1,1,1)        density  2  add.gen.  1/2,1/2,0
SL  3  I(1,1,1)        density  2  add.gen.  1/2,1/2,1/2
SL  5  C(1,1,1/2)      density  4  add.gen.  0,0,1/2   1/2,1/2,0
SL  7  P(1/2,1/2,1)    density  4  add.gen.  1/2,0,0   0,1/2,0
SL 12  P(1/2,1/2,1/2)  density  8  add.gen.  1/2,0,0   0,1/2,0   0,0,1/2
SL 14  C(1/2,1/2,1)    density  8  add.gen.  1/2,0,0   1/4,1/4,0
```

Wyckoff letter   non-characteristic crystallographic orbits

```
8 g 1     SL 0  x,y,0         P4/mbm (8i,j)   x,y,1/2       P4/mbm (8i,j)
                x,x,z         P4/mbm (8k)     x,y,1/4       P4/mnc (8h)
                x,1/2+x,z     P4/nmm (8j)     x,0,z         P4/nmm (8i)
          SL 1  x,x,1/4       P4/mbm (4g,h)
          SL 2  x,1/2+x,1/2   P4/mmm (4l-o)   x,1/2+x,0     P4/mmm (4l-o)
                x,0,1/2       P4/mmm (4j,k)   x,0,0         P4/mmm (4j,k)
          SL 3  x,1/2-x,1/4   I4/mmm (8h)     x,1/2,1/4     I4/mmm (8i,j)
                x,0,1/4       I4/mmm (8i,j)
          SL 7  1/4,1/4,z     P4/mmm (2g,h)
          SL12  1/4,1/4,1/4   P4/mmm (1a-d)
          SL14  1/4,0,1/2     P4/mmm (1a-d)   1/4,0,0       P4/mmm (1a-d)
4 f 2     SL 0  x,x,1/2       P4/mbm (4g,h)
          SL 7  1/4,1/4,1/2   P4/mmm (1a-d)
4 e 2     SL 0  x,x,0         P4/mbm (4g,h)
          SL 7  1/4,1/4,0     P4/mmm (1a-d)
4 d 2     SL 2  0,0,z         P4/mmm (2g,h)
          SL 5  0,0,1/4       P4/mmm (1a-d)
2 c 4     SL 0  0,1/2,z       P4/nmm (2c)
          SL 2  0,1/2,1/2     P4/mmm (1a-d)   0,1/2,0       P4/mmm (1a-d)
          SL 3  0,1/2,3/4     I4/mmm (2a,b)   0,1/2,1/4     I4/mmm (2a,b)
2 b 222   SL 2  0,0,1/2       P4/mmm (1a-d)
2 a 222   SL 2  0,0,0         P4/mmm (1a-d)
```

Space group No. 91 P4$_1$22

Superlattices

SL	1	P(1,1,1/2)	density 2	add.gen.	0,0,1/2	
SL	2	C(1,1,1)	density 2	add.gen.	1/2,1/2,0	
SL	3	I(1,1,1)	density 2	add.gen.	1/2,1/2,1/2	
SL	4	P(1,1,1/4)	density 4	add.gen.	0,0,1/4	
SL	6	I(1,1,1/2)	density 4	add.gen.	1/2,1/2,1/4	
SL	9	P(1,1,1/8)	density 8	add.gen.	0,0,1/8	
SL	10	C(1,1,1/4)	density 8	add.gen.	0,0,1/4	1/2,1/2,0
SL	13	F(1,1,1/2)	density 8	add.gen.	1/2,1/2,0	1/2,0,1/4

Wyckoff letter   non-characteristic crystallographic orbits

Wyckoff letter						
8 d 1	SL 1	x,1/2,0	P4$_2$/mmc (4j-m)	x,0,0	P4$_2$/mmc (4j-m)	
		x,x,1/8	P4$_2$/mcm (4i,j)			
	SL 2	x,1/2-x,1/8	P4$_1$22 (4a,b)			
	SL 3	1/4,1/4,z	I4$_1$/amd (8e)	x,1/2+x,1/8	I4$_1$22 (8d,e)	
	SL 4	0,0,z	P4/mmm (2g,h)	1/2,1/2,z	P4/mmm (2g,h)	
	SL 6	1/2,0,z	I4/mmm (4e)			
	SL 9	1/2,1/2,1/16	P4/mmm (1a-d)	0,0,1/16	P4/mmm (1a-d)	
	SL 10	1/2,0,1/8	P4/mmm (1a-d)			
	SL 13	1/4,1/4,1/8	I4/mmm (2a,b)			
4 c 2	SL 3	1/4,1/4,3/8	I4$_1$/amd (4a,b)	3/4,3/4,3/8	I4$_1$/amd (4a,b)	
	SL 4	0,0,3/8	P4/mmm (1a-d)	1/2,1/2,3/8	P4/mmm (1a-d)	
4 b 2	SL 4	1/2,1/2,0	P4/mmm (1a-d)			
	SL 6	1/2,0,0	I4/mmm (2a,b)			
4 a 2	SL 4	0,0,0	P4/mmm (1a-d)			
	SL 6	0,1/2,0	I4/mmm (2a,b)			

Space group   No. 92   P4$_1$2$_1$2

Superlattices

```
SL  1  P(1,1,1/2)      density  2  add.gen.  0,0,1/2
SL  2  C(1,1,1)        density  2  add.gen.  1/2,1/2,Q
SL  3  I(1,1,1)        density  2  add.gen.  1/2,1/2,1/2
SL  4  P(1,1,1/4)      density  4  add.gen.  0,0,1/4
SL  6  I(1,1,1/2)      density  4  add.gen.  1/2,1/2,1/4
SL  8  F(1,1,1)        density  4  add.gen.  0,1/2,1/2  1/2,0,1/2
SL 10  C(1,1,1/4)      density  8  add.gen.  0,0,1/4  1/2,1/2,0
SL 11  I(1,1,1/4)      density  8  add.gen.  1/2,1/2,1/8
SL 13  F(1,1,1/2)      density  8  add.gen.  1/2,1/2,0  1/2,0,1/4
```

Wyckoff letter   non-characteristic crystallographic orbits

8 b 1	SL 1	x,-x,0	P4$_2$/mnm (4f,g)			
	SL 2	0,y,1/8	P4$_1$22 (4c)	x,1/2-x,1/4	P4$_1$22 (4a,b)	
	SL 3	1/4,1/4,z	I4$_1$/amd (8e)	x,1/2-x,0	I4$_1$22 (8d,e)	
		x,0,1/8	I4$_1$22 (8f)	1/4,0,1/8	I4$_1$/amd (8c,d)	
		3/4,0,1/8	I4$_1$/amd (8c,d)			
	SL 4	1/2,0,z	P4/nmm (2c)			
	SL 6	0,0,z	I4/mmm (4e)			
	SL 8	0,1/4,1/8	I4$_1$/amd (4a,b)	0,3/4,1/8	I4$_1$/amd (4a,b)	
	SL10	1/2,0,1/8	P4/mmm (1a-d)	1/2,0,0	P4/mmm (1a-d)	
		0,0,1/8	P4/mmm (1a-d)			
	SL11	1/2,0,1/16	I4/mmm (2a,b)	1/2,0,3/16	I4/mmm (2a,b)	
	SL13	1/4,1/4,1/4	I4/mmm (2a,b)			
4 a 2	SL 3	3/4,3/4,0	I4$_1$/amd (4a,b)	1/4,1/4,0	I4$_1$/amd (4a,b)	
	SL 6	0,0,0	I4/mmm (2a,b)	1/2,1/2,0	I4/mmm (2a,b)	

Space group   No. 93   P4$_2$22

Superlattices

```
SL   0   P(1,1,1)         density  1  add.gen.
SL   1   P(1,1,1/2)       density  2  add.gen.  0,0,1/2
SL   2   C(1,1,1)         density  2  add.gen.  1/2,1/2,0
SL   3   I(1,1,1)         density  2  add.gen.  1/2,1/2,1/2
SL   4   P(1,1,1/4)       density  4  add.gen.  0,0,1/4
SL   5   C(1,1,1/2)       density  4  add.gen.  0,0,1/2  1/2,1/2,0
SL   8   F(1,1,1)         density  4  add.gen.  0,1/2,1/2  1/2,0,1/2
SL  12   P(1/2,1/2,1/2)   density  8  add.gen.  1/2,0,0  0,1/2,0  0,0,1/2
```

Wyckoff letter   non-characteristic crystallographic orbits

8 p 1	SL 0	x,y,0	P4$_2$/mmc (8q)	x,1/2,z	P4$_2$/mmc (8o,p)
		x,0,z	P4$_2$/mmc (8o,p)	x,x,z	P4$_2$/mcm (8o)
		x,y,1/4	P4$_2$/mcm (8n)	x,1/2+x,z	P4$_2$/nnm (8m)
	SL 1	x,1/2,1/4	P4/mmm (4l-o)	x,0,1/4	P4/mmm (4l-o)
		x,x,0	P4/mmm (4j,k)		
	SL 2	x,1/2+x,1/4	P4$_2$/mmc (4j-m)	x,1/2-x,1/4	P4$_2$/mmc (4j-m)
	SL 3	x,1/2-x,0	I4/mmm (8h)		
	SL 8	1/4,1/4,z	I4/mmm (4e)		
	SL 12	1/4,1/4,0	P4/mmm (1a-d)		
4 o 2	SL 0	x,x,3/4	P4$_2$/mcm (4i,j)		
	SL 8	1/4,1/4,3/4	I4/mmm (2a,b)		
4 n 2	SL 0	x,x,1/4	P4$_2$/mcm (4i,j)		
	SL 8	1/4,1/4,1/4	I4/mmm (2a,b)		
4 m 2	SL 0	x,1/2,0	P4$_2$/mmc (4j-m)		
4 l 2	SL 0	x,0,1/2	P4$_2$/mmc (4j-m)		
4 k 2	SL 0	x,1/2,1/2	P4$_2$/mmc (4j-m)		
4 j 2	SL 0	x,0,0	P4$_2$/mmc (4j-m)		
4 i 2	SL 3	0,1/2,z	I4/mmm (4e)		
	SL 5	0,1/2,1/4	P4/mmm (1a-d)		
4 h 2	SL 1	1/2,1/2,z	P4/mmm (2g,h)		
	SL 4	1/2,1/2,1/8	P4/mmm (1a-d)		
4 g 2	SL 1	0,0,z	P4/mmm (2g,h)		
	SL 4	0,0,1/8	P4/mmm (1a-d)		
2 f 222	SL 1	1/2,1/2,1/4	P4/mmm (1a-d)		
2 e 222	SL 1	0,0,1/4	P4/mmm (1a-d)		
2 d 222	SL 3	0,1/2,1/2	I4/mmm (2a,b)		
2 c 222	SL 3	0,1/2,0	I4/mmm (2a,b)		
2 b 222	SL 1	1/2,1/2,0	P4/mmm (1a-d)		
2 a 222	SL 1	0,0,0	P4/mmm (1a-d)		

Space group No. 94 P4₂2₁2

Superlattices

SL	0	P(1,1,1)	density	1	add.gen.			
SL	1	P(1,1,1/2)	density	2	add.gen.	0,0,1/2		
SL	2	C(1,1,1)	density	2	add.gen.	1/2,1/2,0		
SL	3	I(1,1,1)	density	2	add.gen.	1/2,1/2,1/2		
SL	5	C(1,1,1/2)	density	4	add.gen.	0,0,1/2	1/2,1/2,0	
SL	6	I(1,1,1/2)	density	4	add.gen.	1/2,1/2,1/4		
SL	8	F(1,1,1)	density	4	add.gen.	0,1/2,1/2	1/2,0,1/2	
SL	12	P(1/2,1/2,1/2)	density	8	add.gen.	1/2,0,0	0,1/2,0	0,0,1/2
SL	15	I(1/2,1/2,1)	density	8	add.gen.	1/2,0,0	1/4,1/4,1/2	

Wyckoff letter   non-characteristic crystallographic orbits

Wyckoff						
8 g 1	SL 0	x,y,1/4	P4₂/mbc (8h)	x,x,z	P4₂/mnm (8j)
		x,y,0	P4₂/mnm (8i)	x,0,z	P4₂/nmc (8g)
		x,1/2+x,z	P4₂/ncm (8i)		
	SL 1	x,x,1/4	P4/mbm (4g,h)		
	SL 2	x,1/2+x,1/2	P4₂/mmc (4j-m)	x,1/2+x,0	P4₂/mmc (4j-m)
		x,0,3/4	P4₂/mcm (4i,j)	x,0,1/4	P4₂/mcm (4i,j)
	SL 3	x,1/2+x,1/4	I4/mcm (8h)	x,0,1/2	I4/mmm (8i,j)
		x,0,0	I4/mmm (8i,j)		
	SL 8	1/4,1/4,z	I4/mmm (4e)		
	SL12	1/4,1/4,1/4	P4/mmm (1a-d)		
	SL15	1/4,0,3/4	I4/mmm (2a,b)	1/4,0,1/4	I4/mmm (2a,b)
4 f 2	SL 0	x,x,1/2	P4₂/mnm (4f,g)		
	SL 8	1/4,1/4,1/2	I4/mmm (2a,b)		
4 e 2	SL 0	x,x,0	P4₂/mnm (4f,g)		
	SL 8	1/4,1/4,0	I4/mmm (2a,b)		
4 d 2	SL 1	0,1/2,z	P4/nmm (2c)		
	SL 5	0,1/2,0	P4/mmm (1a-d)	0,1/2,1/4	P4/mmm (1a-d)
	SL 6	0,1/2,1/8	I4/mmm (2a,b)	0,1/2,3/8	I4/mmm (2a,b)
4 c 2	SL 3	0,0,z	I4/mmm (4e)		
	SL 5	0,0,1/4	P4/mmm (1a-d)		
2 b 222	SL 3	0,0,1/2	I4/mmm (2a,b)		
2 a 222	SL 3	0,0,0	I4/mmm (2a,b)		

Space group No. 95 $P4_322$

Superlattices

```
SL  1  P(1,1,1/2)      density  2  add.gen.  0,0,1/2
SL  2  C(1,1,1)        density  2  add.gen.  1/2,1/2,0
SL  3  I(1,1,1)        density  2  add.gen.  1/2,1/2,1/2
SL  4  P(1,1,1/4)      density  4  add.gen.  0,0,1/4
SL  6  I(1,1,1/2)      density  4  add.gen.  1/2,1/2,1/4
SL  9  P(1,1,1/8)      density  8  add.gen.  0,0,1/8
SL 10  C(1,1,1/4)      density  8  add.gen.  0,0,1/4  1/2,1/2,0
SL 13  F(1,1,1/2)      density  8  add.gen.  1/2,1/2,0  1/2,0,1/4
```

Wyckoff letter   non-characteristic crystallographic orbits

8 d 1	SL 1	x,1/2,0	$P4_2$/mmc (4j-m)	x,0,0	$P4_2$/mmc (4j-m)
		x,x,3/8	$P4_2$/mcm (4i,j)		
	SL 2	x,1/2+x,1/8	$P4_3$22 (4a,b)		
	SL 3	1/4,1/4,z	$I4_1$/amd (8e)	x,1/2-x.1/8	$I4_1$22 (8d,e)
	SL 4	0,0,z	P4/mmm (2g,h)	1/2,1/2,z	P4/mmm (2g,h)
	SL 6	1/2,0,z	I4/mmm (4e)		
	SL 9	0,0,1/16	P4/mmm (1a-d)	1/2,1/2,1/16	P4/mmm (1a-d)
	SL10	1/2,0,1/8	I4/mmm (2a,b)		
	SL13	1/4,1/4,3/8			
4 c 2	SL 3	1/4,1/4,5/8	$I4_1$/amd (4a,b)	3/4,3/4,5/8	$I4_1$/amd (4a,b)
	SL 4	0,0,5/8	P4/mmm (1a-d)	1/2,1/2,5/8	P4/mmm (1a-d)
4 b 2	SL 4	1/2,1/2,0	P4/mmm (1a-d)		
	SL 6	1/2,0,0	I4/mmm (2a,b)		
4 a 2	SL 4	0,0,0	P4/mmm (1a-d)		
	SL 6	0,1/2,0	I4/mmm (2a,b)		

Space group   No. 96   P4$_3$2$_1$2

Superlattices

```
SL  1  P(1,1,1/2)      density  2  add.gen.  0,0,1/2
SL  2  C(1,1,1)        density  2  add.gen.  1/2,1/2,0
SL  3  I(1,1,1)        density  2  add.gen.  1/2,1/2,1/2
SL  4  P(1,1,1/4)      density  4  add.gen.  0,0,1/4
SL  6  I(1,1,1/2)      density  4  add.gen.  1/2,1/2,1/4
SL  8  F(1,1,1)        density  4  add.gen.  0,1/2,1/2  1/2,0,1/2
SL 10  C(1,1,1/4)      density  8  add.gen.  0,0,1/4  1/2,1/2,0
SL 11  I(1,1,1/4)      density  8  add.gen.  1/2,1/2,1/8
SL 13  F(1,1,1/2)      density  8  add.gen.  1/2,1/2,0  1/2,0,1/4
```

Wyckoff letter   non-characteristic crystallographic orbits

```
   8 b 1     SL 1  x,-x,0          P4₂/mnm (4f,g)
             SL 2  x,1/2-x,1/4     P4₃22 (4a,b)      x,0,1/8        P4₃22 (4c)
             SL 3  1/4,1/4,z       I4₁/amd (8e)      0,y,1/8        I4₁22 (8f)
                   0,1/4,1/8       I4₁/amd (8c,d)    0,3/4,1/8      I4₁/amd (8c,d)
                   x,1/2-x,0       I4₁22 (8d,e)
             SL 4  1/2,0,z         P4/nmm (2c)
             SL 6  0,0,z           I4/mmm (4e)
             SL 8  0,1/4,3/8       I4₁/amd (4a,b)    0,1/4,7/8      I4₁/amd (4a,b)
             SL10  0,0,1/8         P4/mmm (1a-d)     1/2,0,0        P4/mmm (1a-d)
                   1/2,0,1/8       P4/mmm (1a-d)
             SL11  1/2,0,3/16      I4/mmm (2a,b)     1/2,0,1/16     I4/mmm (2a,b)
             SL13  1/4,1/4,1/4     I4/mmm (2a,b)
   4 a 2     SL 3  3/4,3/4,0       I4₁/amd (4a,b)    1/4,1/4,0      I4₁/amd (4a,b)
             SL 6  0,0,0           I4/mmm (2a,b)     1/2,1/2,0      I4/mmm (2a,b)
```

Space group   No. 97   I422

Superlattices

```
SL  0  I(1,1,1)        density  1  add.gen.
SL  1  C(1,1,1/2)      density  2  add.gen.  0,0,1/2
SL  2  C(1,1,1/4)      density  4  add.gen.  0,0,1/4
SL  3  P(1/2,1/2,1/2)  density  4  add.gen.  1/2,0,0   0,1/2,0
SL  6  P(1/2,1/2,1/4)  density  8  add.gen.  1/2,0,0   0,0,1/4
SL  8  C(1/2,1/2,1/2)  density  8  add.gen.  1/2,0,0   1/4,1/4,0
```

Wyckoff letter   non-characteristic crystallographic orbits

```
16 k 1    SL 0   x,0,z         I4/mmm (16n)   x,x,z        I4/mmm (16m)
                 x,y,0         I4/mmm (16l)   x,1/2+x,z    I4/mcm (16l)
                 x,y,1/4       I4/mcm (16k)
          SL 1   x,x,1/4       P4/mmm (4l-o)  x,0,1/4      P4/mmm (4j,k)
                 x,1/2+x,0     P4/mmm (4l-o)
          SL 3   1/4,1/4,z     P4/mmm (2g,h)
          SL 6   1/4,1/4,1/8   P4/mmm (1a-d)
          SL 8   1/4,0,1/4     P4/mmm (1a-d)
 8 j 2    SL 0   x,1/2+x,1/4   I4/mcm (8h)
          SL 3   1/4,3/4,1/4   P4/mmm (1a-d)
 8 i 2    SL 0   x,0,1/2       I4/mmm (8i,j)
 8 h 2    SL 0   x,0,0         I4/mmm (8i,j)
 8 g 2    SL 0   x,x,0         I4/mmm (8h)
          SL 3   1/4,1/4,0     P4/mmm (1a-d)
 8 f 2    SL 1   0,1/2,z       P4/mmm (2g,h)
          SL 2   0,1/2,1/8     P4/mmm (1a-d)
 4 e 4    SL 0   0,0,z         I4/mmm (4e)
          SL 1   0,0,1/4       P4/mmm (1a-d)
 4 d 222  SL 1   0,1/2,1/4     P4/mmm (1a-d)
 4 c 222  SL 1   0,1/2,0       P4/mmm (1a-d)
 2 b 422  SL 0   0,0,1/2       I4/mmm (2a,b)
 2 a 422  SL 0   0,0,0         I4/mmm (2a,b)
```

Space group   No. 98 $I4_122$

Superlattices

```
SL  0  I(1,1,1)          density  1  add.gen.
SL  1  C(1,1,1/2)        density  2  add.gen.  ,0,0,1/2
SL  2  C(1,1,1/4)        density  4  add.gen.  0,0,1/4
SL  4  F(1,1,1/2)        density  4  add.gen.  0,1/2,1/4
SL  5  C(1,1,1/8)        density  8  add.gen.  0,0,1/8
SL  9  I(1/2,1/2,1/2)    density  8  add.gen.  1/2,0,0   0,1/2,0
```

Wyckoff letter   non-characteristic crystallographic orbits

```
 16 g 1      SL 0  x,0,z           I4₁/amd (16h)
             SL 1  x,1/2+x,0       P4₂/mmc (4j-m)   x,1/2-x,0      P4₂/mmc (4j-m)
                   1/4,y,1/8       P4₂/mcm (4i,j)
             SL 2  3/4,1/4,z       P4/mmm (2g,h)    1/4,1/4,z      P4/mmm (2g,h)
             SL 5  3/4,1/4,1/16 P4/mmm (1a-d)       1/4,1/4,1/16 P4/mmm (1a-d)
             SL 9  1/4,0,1/8       I4/mmm ,(2a,b)
  8 f 2      SL 0  1/2,1/4,1/8     I4₁/amd (8c,d)   0,1/4,1/8      I4₁/amd (8c,d)
             SL 2  3/4,1/4,1/8     P4/mmm (1a-d)    1/4,1/4,1/8    P4/mmm (1a-d)
  8 e 2      SL 2  3/4,1/4,0       P4/mmm (1a-d)
  8 d 2      SL 2  1/4,1/4,0       P4/mmm (1a-d)
  8 c 2      SL 0  0,0,z           I4₁/amd (8e)
             SL 4  0,0,1/4         I4/mmm (2a,b)
  4 b 222    SL 0  0,0,1/2         I4₁/amd (4a,b)
  4 a 222    SL 0  0,0,0           I4₁/amd (4a,b)
```

Space group No. 99 P4mm

Superlattices

```
SL  0  P(1,1,1)        density 1  add.gen.
SL  2  C(1,1,1)        density 2  add.gen.  1/2,1/2,0
SL  7  P(1/2,1/2,1)    density 4  add.gen.  1/2,0,0  0,1/2,0
```

Wyckoff letter   non-characteristic crystallographic orbits

```
 8 g 1     SL 0   x,y,z          P4/mmm (8p,q)
           SL 2   x,1/2+x,z      P4/mmm (4l-o)
 4 f m     SL 0   x,1/2,z        P4/mmm (4l-o)
 4 e m     SL 0   x,0,z          P4/mmm (4l-o)
 4 d m     SL 0   x,x,z          P4/mmm (4j,k)
           SL 7   1/4,1/4,z      P4/mmm (1a-d)
 2 c mm2   SL 2   1/2,0,z        P4/mmm (1a-d)
 1 b 4mm   SL 0   1/2,1/2,z      P4/mmm (1a-d)
 1 a 4mm   SL 0   0,0,z          P4/mmm (1a-d)
```

Space group No.100 P4bm

Superlattices

```
SL  0  P(1,1,1)        density 1  add.gen.
SL  2  C(1,1,1)        density 2  add.gen.  1/2,1/2,0
SL  7  P(1/2,1/2,1)    density 4  add.gen.  1/2,0,0  0,1/2,0
SL 14  C(1/2,1/2,1)    density 8  add.gen.  1/2,0,0  1/4,1/4,0
```

Wyckoff letter   non-characteristic crystallographic orbits

```
 8 d 1     SL 0   x,y,z          P4/mbm (8i,j)
           SL 2   x,x,z          P4/mmm (4l-o)   x,0,z      P4/mmm (4j,k)
           SL14   1/4,0,z        P4/mmm (1a-d)
 4 c m     SL 0   x,1/2+x,z      P4/mbm (4g,h)
           SL 7   1/4,3/4,z      P4/mmm (1a-d)
 2 b mm2   SL 2   1/2,0,z        P4/mmm (1a-d)
 2 a 4     SL 2   0,0,z          P4/mmm (1a-d)
```

Space group No.101 P4₂cm

Superlattices

```
SL  0  P(1,1,1)        density  1  add.gen.
SL  1  P(1,1,1/2)      density  2  add.gen.  0,0,1/2
SL  2  C(1,1,1)        density  2  add.gen.  1/2,1/2,0
SL  5  C(1,1,1/2)      density  4  add.gen.  0,0,1/2  1/2,1/2,0
SL  8  F(1,1,1)        density  4  add.gen.  0,1/2,1/2  1/2,0,1/2
```

Wyckoff letter   non-characteristic crystallographic orbits

```
8 e 1      SL 0   x,y,z          P4₂/mcm (8n)
           SL 1   x,1/2,z        P4/mmm (4l-o)   x,0,z       P4/mmm (4l-o)
           SL 2   x,1/2+x,z      P4₂/mmc (4j-m)
4 d m      SL 0   x,x,z          P4₂/mcm (4i,j)
           SL 8   1/4,1/4,z      I4/mmm (2a,b)
4 c 2      SL 5   0,1/2,z        P4/mmm (1a-d)
2 b mm2    SL 1   1/2,1/2,z      P4/mmm (1a-d)
2 a mm2    SL 1   0,0,z          P4/mmm (1a-d)
```

Space group   No.102 P4₂nm

Superlattices

```
SL  0  P(1,1,1)        density  1  add.gen.
SL  2  C(1,1,1)        density  2  add.gen.  1/2,1/2,0
SL  3  I(1,1,1)        density  2  add.gen.  1/2,1/2,1/2
SL  5  C(1,1,1/2)      density  4  add.gen.  0,0,1/2  1/2,1/2,0
SL  8  F(1,1,1)        density  4  add.gen.  0,1/2,1/2  1/2,0,1/2
```

Wyckoff letter   non-characteristic crystallographic orbits

```
8 d 1      SL 0   x,y,z          P4₂/mnm (8i)
           SL 2   x,1/2+x,z      P4₂/mmc (4j-m)
           SL 3   x,0,z          I4/mmm (8i,j)
4 c m      SL 0   x,x,z          P4₂/mnm (4f,g)
           SL 8   1/4,1/4,z      I4/mmm (2a,b)
4 b 2      SL 5   0,1/2,z        P4/mmm (1a-d)
2 a mm2    SL 3   0,0,z          I4/mmm (2a,b)
```

Space group No.103 P4cc

Superlattices

```
SL  0  P(1,1,1)        density  1  add.gen.
SL  1  P(1,1,1/2)      density  2  add.gen.  0,0,1/2
SL  3  I(1,1,1)        density  2  add.gen.  1/2,1/2,1/2
SL  5  C(1,1,1/2)      density  4  add.gen.  0,0,1/2  1/2,1/2,0
SL 12  P(1/2,1/2,1/2)  density  8  add.gen.  1/2,0,0  0,1/2,0  0,0,1/2
```

Wyckoff letter   non-characteristic crystallographic orbits

```
   8 d 1    SL  0  x,y,z         P4/mcc (8m)
            SL  1  x,x,z         P4/mmm (4j,k)     x,1/2,z      P4/mmm (4l-o)
                   x,0,z         P4/mmm (4l-o)
            SL  3  x,1/2+x,z     I4/mmm (8h)
            SL 12  1/4,1/4,z     P4/mmm (1a-d)
   4 c 2    SL  5  0,1/2,z       P4/mmm (1a-d)
   2 b 4    SL  1  1/2,1/2,z     P4/mmm (1a-d)
   2 a 4    SL  1  0,0,z         P4/mmm (1a-d)
```


Space group No.104 P4nc

Superlattices

```
SI  0  P(1,1,1)        density  1  add.gen.
SL  1  P(1,1,1/2)      density  2  add.gen.  0,0,1/2
SL  3  I(1,1,1)        density  2  add.gen.  1/2,1/2,1/2
SL  5  C(1,1,1/2)      density  4  add.gen.  0,0,1/2  1/2,1/2,0
SL 12  P(1/2,1/2,1/2)  density  8  add.gen.  1/2,0,0  0,1/2,0  0,0,1/2
```

Wyckoff letter   non-characteristic crystallographic orbits

```
   8 c 1    SL  0  x,y,z         P4/mnc (8h)
            SL  1  x,1/2+x,z     P4/mbm (4g,h)
            SL  3  x,x,z         I4/mmm (8h)       x,0,z        I4/mmm (8i,j)
            SL 12  1/4,1/4,z     P4/mmm (1a-d)
   4 b 2    SL  5  0,1/2,z       P4/mmm (1a-d)
   2 a 4    SL  3  0,0,z         I4/mmm (2a,b)
```

Space group   No.105   P4$_2$mc

Superlattices

```
SL  0   P(1,1,1)           density  1   add.gen.
SL  1   P(1,1,1/2)         density  2   add.gen.   0,0,1/2
SL  3   I(1,1,1)           density  2   add.gen.   1/2,1/2,1/2
SL 12   P(1/2,1/2,1/2)     density  8   add.gen.   1/2,0,0   0,1/2,0   0,0,1/2
```

Wyckoff letter   non-characteristic crystallographic orbits

```
8 f 1      SL 0    x,y,z         P4₂/mmc (8q)
           SL 1    x,x,z         P4/mmm  (4j,k)
           SL 3    x,1/2+x,z     I4/mmm  (8h)
           SL12    1/4,1/4,z     P4/mmm  (1a-d)
4 e m      SL 0    x,1/2,z       P4₂/mmc (4j-m)
4 d m      SL 0    x,0,z         P4₂/mmc (4j-m)
2 c mm2    SL 3    0,1/2,z       I4/mmm  (2a,b)
2 b mm2    SL 1    1/2,1/2,z     P4/mmm  (1a-d)
2 a mm2    SL 1    0,0,z         P4/mmm  (1a-d)
```

Space group   No.106   P4$_2$bc

Superlattices

```
SL  0   P(1,1,1)           density  1   add.gen.
SL  1   P(1,1,1/2)         density  2   add.gen.   0,0,1/2
SL  2   C(1,1,1)           density  2   add.gen.   1/2,1/2,0
SL  3   I(1,1,1)           density  2   add.gen.   1/2,1/2,1/2
SL  5   C(1,1,1/2)         density  4   add.gen.   0,0,1/2   1/2,1/2,0
SL 12   P(1/2,1/2,1/2)     density  8   add.gen.   1/2,0,0   0,1/2,0   0,0,1/2
SL 15   I(1/2,1/2,1)       density  8   add.gen.   1/2,0,0   1/4,1/4,1/2
```

Wyckoff letter   non-characteristic crystallographic orbits

```
8 c 1      SL 0    x,y,z         P4₂/mbc (8h)
           SL 1    x,1/2+x,z     P4/mbm  (4g,h)
           SL 2    x,0,z         P4₂/mcm (4i,j)
           SL 3    x,x,z         I4/mcm  (8h)
           SL12    1/4,1/4,z     P4/mmm  (1a-d)
           SL15    1/4,0,z       I4/mmm  (2a,b)
4 b 2      SL 5    0,1/2,z       P4/mmm  (1a-d)
4 a 2      SL 5    0,0,z         P4/mmm  (1a-d)
```

Space group No.107 I4mm

Superlattices

```
SL  0  I(1,1,1)          density  1  add.gen.
SL  1  C(1,1,1/2)        density  2  add.gen.  0,0,1/2
SL  3  P(1/2,1/2,1/2)    density  4  add.gen.  1/2,0,0  0,1/2,0
```

Wyckoff letter   non-characteristic crystallographic orbits

```
16 e 1      SL 0   x,y,z        I4/mmm (16l)
            SL 1   x,1/2+x,z    P4/mmm (4l-o)
 8 d m      SL 0   x,0,z        I4/mmm (8i,j)
 8 c m      SL 0   x,x,z        I4/mmm (8h)
            SL 3   1/4,1/4,z    P4/mmm (1a-d)
 4 b mm2    SL 1   0,1/2,z      P4/mmm (1a-d)
 2 a 4mm    SL 0   0,0,z        I4/mmm (2a,b)
```

Space group No.108 I4cm

Superlattices

```
SL  0  I(1,1,1)          density  1  add.gen.
SL  1  C(1,1,1/2)        density  2  add.gen.  0,0,1/2
SL  3  P(1/2,1/2,1/2)    density  4  add.gen.  1/2,0,0  0,1/2,0
SL  8  C(1/2,1/2,1/2)    density  8  add.gen.  1/2,0,0  1/4,1/4,0
```

Wyckoff letter   non-characteristic crystallographic orbits

```
16 d 1      SL 0   x,y,z        I4/mcm (16k)
            SL 1   x,0,z        P4/mmm (4j,k)   x,x,z        P4/mmm (4l-o)
            SL 8   1/4,0,z      P4/mmm (1a-d)
 8 c m      SL 0   x,1/2+x,z    I4/mcm (8h)
            SL 3   1/4,3/4,z    P4/mmm (1a-d)
 4 b mm2    SL 1   1/2,0,z      P4/mmm (1a-d)
 4 a 4      SL 1   0,0,z        P4/mmm (1a-d)
```

Space group No.109 I4₁md

Superlattices

```
SL  0  I(1,1,1)           density  1  add.gen.
SL  6  P(1/2,1/2,1/4)     density  4  add.gen.   1/2,0,0   0,0,1/4
```

Wyckoff letter   non-characteristic crystallographic orbits

```
  16 c 1      SL 0   x,x,z        I4₁/amd (16g)    x,1/4,z      I4₁/amd (16f)
              SL 6   1/4,1/4,z    P4/mmm (1a-d)
   8 b m      SL 0   0,1/4,z      I4₁/amd (8c,d)
   4 a mm2    SL 0   0,0,z        I4₁/amd (4a,b)
```

Space group No.110 I4₁cd

Superlattices

```
SL  0  I(1,1,1)           density  1  add.gen.
SL  1  C(1,1,1/2)         density  2  add.gen.   0,0,1/2
SL  4  F(1,1,1/2)         density  4  add.gen.   0,1/2,1/4
SL  6  P(1/2,1/2,1/4)     density  8  add.gen.   1/2,0,0   0,0,1/4
SL  9  I(1/2,1/2,1/2)     density  8  add.gen.   1/2,0,0   0,1/2,0   1/4,1/4,1/4
```

Wyckoff letter   non-characteristic crystallographic orbits

```
  16 b 1      SL 0   x,x,z        I4₁/acd (16f)    x,1/4,z      I4₁/acd (16e)
              SL 1   x,0,z        P4₂/mnm (4f,g)
              SL 6   1/4,1/4,z    P4/mmm (1a-d)
              SL 9   1/4,0,z      I4/mmm (2a,b)
   8 a 2      SL 4   0,0,z        I4/mmm (2a,b)
```

Space group No.111 P$\bar{4}$2m

Superlattices

SL	0	P(1,1,1)	density	1	add.gen.		
SL	1	P(1,1,1/2)	density	2	add.gen.	0,0,1/2	
SL	2	C(1,1,1)	density	2	add.gen.	1/2,1/2,0	
SL	5	C(1,1,1/2)	density	4	add.gen.	0,0,1/2	1/2,1/2,0
SL	7	P(1/2,1/2,1)	density	4	add.gen.	1/2,0,0	0,1/2,0
SL	8	F(1,1,1)	density	4	add.gen.	0,1/2,1/2	1/2,0,1/2

Wyckoff letter   non-characteristic crystallographic orbits

Wyckoff	SL	coord	orbit	coord2	orbit2
8 o 1	SL 0	x,y,0	P4/mmm (8p,q)	x,y,1/2	P4/mmm (8p,q)
		x,0,z	P4/mmm (8s,t)	x,1/2,z	P4/mmm (8s,t)
		x,y,1/4	P4$_2$/mcm (8n)		
	SL 1	x,1/2,1/4	P4/mmm (4l-o)	x,0,1/4	P4/mmm (4l-o)
	SL 2	x,1/2+x,z	P$\bar{4}$m2 (4j,k)	x,1/2+x,0	P4/mmm (4l-o)
		x,1/2+x,1/2	P4/mmm (4l-o)	x,1/2+x,1/4	P4$_2$/mmc (4j-m)
		x,1/2+x,3/4	P4$_2$/mmc (4j-m)		
4 n m	SL 0	x,x,0	P4/mmm (4j,k)	x,x,1/2	P4/mmm (4j,k)
		x,x,1/4	P4$_2$/mcm (4i,j)	x,x,3/4	P4$_2$/mcm (4i,j)
	SL 2	1/4,1/4,z	P4/nmm (2c)		
	SL 7	1/4,1/4,1/2	P4/mmm (1a-d)	1/4,1/4,0	P4/mmm (1a-d)
	SL 8	1/4,1/4,1/4	I4/mmm (2a,b)	1/4,1/4,3/4	I4/mmm (2a,b)
4 m 2	SL 2	0,1/2,z	P4/mmm (2g,h)		
	SL 5	0,1/2,1/4	P4/mmm (1a-d)		
4 l 2	SL 0	x,1/2,0	P4/mmm (4l-o)		
4 k 2	SL 0	x,0,1/2	P4/mmm (4l-o)		
4 j 2	SL 0	x,1/2,1/2	P4/mmm (4l-o)		
4 ı 2	SL 0	x,0,0	P4/mmm (4l-o)		
2 h mm2	SL 0	1/2,1/2,z	P4/mmm (2g,h)		
	SL 1	1/2,1/2,1/4	P4/mmm (1a-d)		
2 g mm2	SL 0	0,0,z	P4/mmm (2g,h)		
	SL 1	0,0,1/4	P4/mmm (1a-d)		
2 f 222	SL 2	1/2,0,1/2	P4/mmm (1a-d)		
2 e 222	SL 2	1/2,0,0	P4/mmm (1a-d)		
1 d $\bar{4}$2m	SL 0	1/2,1/2,0	P4/mmm (1a-d)		
1 c $\bar{4}$2m	SL 0	0,0,1/2	P4/mmm (1a-d)		
1 b $\bar{4}$2m	SL 0	1/2,1/2,1/2	P4/mmm (1a-d)		
1 a $\bar{4}$2m	SL 0	0,0,0	P4/mmm (1a-d)		

Space group   No.112   P4̄2c

Superlattices

```
SL  0  P(1,1,1)        density 1  add.gen.
SL  1  P(1,1,1/2)      density 2  add.gen.  0,0,1/2
SL  3  I(1,1,1)        density 2  add.gen.  1/2,1/2,1/2
SL  4  P(1,1,1/4)      density 4  add.gen.  0,0,1/4
SL  5  C(1,1,1/2)      density 4  add.gen.  0,0,1/2   1/2,1/2,0
SL 12  P(1/2,1/2,1/2)  density 8  add.gen.  1/2,0,0  0,1/2,0  0,0,1/2
SL 13  F(1,1,1/2)      density 8  add.gen.  1/2,1/2,0  1/2,0,1/4
```

Wyckoff letter   non-characteristic crystallographic orbits

8 n 1	SL 0	x,y,0	P4/mcc (8m)	x,y,1/4	P4₂/mmc (8q)
		x,1/2,z	P4₂/mmc (8o,p)	x,0,z	P4₂/mmc (8o,p)
	SL 1	x,x,z	P4̄2m (4n)	x,x,0	P4/mmm (4j,k)
		x,x,1/4	P4/mmm (4j,k)	x,x,1/8	P4₂/mcm (4i,j)
		x,x,3/8	P4₂/mcm (4i,j)	x,1/2,0	P4/mmm (4l-o)
		x,0,0	P4/mmm (4l-o)		
	SL 3	x,1/2+x,z	I4̄2m (8i)	x,1/2+x,1/4	I4/mmm (8h)
		x,1/2+x,0	I4/mcm (8h)		
	SL 5	1/4,1/4,z	P4/nmm (2c)		
	SL12	1/4,1/4,1/4	P4/mmm (1a-d)	1/4,1/4,0	P4/mmm (1a-d)
	SL13	1/4,1/4,1/8	I4/mmm (2a,b)	1/4,1/4,3/8	I4/mmm (2a,b)
4 m 2	SL 3	0,1/2,z	I4/mmm (4e)		
	SL 5	0,1/2,0	P4/mmm (1a-d)		
4 l 2	SL 1	1/2,1/2,z	P4/mmm (2g,h)		
	SL 4	1/2,1/2,1/8	P4/mmm (1a-d)		
4 k 2	SL 1	0,0,z	P4/mmm (2g,h)		
	SL 4	0,0,1/8	P4/mmm (1a-d)		
4 j 2	SL 0	0,y,1/4	P4₂/mmc (4j-m)		
4 i 2	SL 0	x,1/2,1/4	P4₂/mmc (4j-m)		
4 h 2	SL 0	1/2,y,1/4	P4₂/mmc (4j-m)		
4 g 2	SL 0	x,0,1/4	P4₂/mmc (4j-m)		
2 f 4̄	SL 1	1/2,1/2,0	P4/mmm (1a-d)		
2 e 4̄	SL 1	0,0,0	P4/mmm (1a-d)		
2 d 222	SL 3	0,1/2,1/4	I4/mmm (2a,b)		
2 c 222	SL 1	1/2,1/2,1/4	P4/mmm (1a-d)		
2 b 222	SL 3	1/2,0,1/4	I4/mmm (2a,b)		
2 a 222	SL 1	0,0,1/4	P4/mmm (1a-d)		

Space group No.113 P$\bar{4}2_1$m

Superlattices

SL	0	P(1,1,1)	density	1	add.gen.		
SL	2	C(1,1,1)	density	2	add.gen.	1/2,1/2,0	
SL	3	I(1,1,1)	density	2	add.gen.	1/2,1/2,1/2	
SL	5	C(1,1,1/2)	density	4	add.gen.	0,0,1/2	1/2,1/2,0
SL	7	P(1/2,1/2,1)	density	4	add.gen.	1/2,0,0	0,1/2,0
SL	8	F(1,1,1)	density	4	add.gen.	0,1/2,1/2	1/2,0,1/2
SL	14	C(1/2,1/2,1)	density	8	add.gen.	1/2,0,0	1/4.1/4,0

Wyckoff letter		non-characteristic crystallographic orbits			
8 f 1	SL 0	x,y,1/2	P4/mbm (8i,j)	x,y,0	P4/mbm (8i,j)
		x,0,z	P4/nmm (8i)	x,y,1/4	P4$_2$/mnm (8i)
	SL 2	x,x,z	P$\bar{4}$m2 (4j,k)	x,x,0	P4/mmm (4l-o)
		x,x,1/2	P4/mmm (4l-o)	x,x,1/4	P4$_2$/mmc (4j-m)
		x,x,3/4	P4$_2$/mmc (4j-m)	x,0,1/2	P4/mmm (4j,k)
		x,0,0	P4/mmm (4j,k)		
	SL 3	x,0,3/4	I4/mmm (8i,j)	x,0,1/4	I4/mmm (8i,j)
	SL14	1/4,0,1/2	P4/mmm (1a-d)	1/4,0,0	P4/mmm (1a-d)
4 e m	SL 0	x,1/2+x,1/2	P4/mbm (4g,h)	x,1/2+x,0	P4/mbm (4g,h)
		x,1/2+x,1/4	P4$_2$/mnm (4f,g)	x,1/2+x,3/4	P4$_2$/mnm (4f,g)
	SL 2	1/4,3/4,z	P4/nmm (2c)		
	SL 7	1/4,3/4,0	P4/mmm (1a-d)	1/4,3/4,1/2	P4/mmm (1a-d)
	SL 8	1/4,3/4,1/4	I4/mmm (2a,b)	1/4,3/4,3/4	I4/mmm (2a,b)
4 d 2	SL 2	0,0,z	P4/mmm (2g,h)		
	SL 5	0,0,1/4	P4/mmm (1a-d)		
2 c mm2	SL 0	0,1/2,z	P4/nmm (2c)		
	SL 2	0,1/2,1/2	P4/mmm (1a-d)	0,1/2,0	P4/mmm (1a-d)
	SL 3	0,1/2,1/4	I4/mmm (2a,b)	0,1/2,3/4	I4/mmm (2a,b)
2 b $\bar{4}$	SL 2	0,0,1/2	P4/mmm (1a-d)		
2 a $\bar{4}$	SL 2	0,0,0	P4/mmm (1a-d)		

Space group No.114 P$\bar{4}2_1$c

Superlattices

SL	0	P(1,1,1)	density	1	add.gen.		
SL	1	P(1,1,1/2)	density	2	add.gen.	0,0,1/2	
SL	2	C(1,1,1)	density	2	add.gen.	1/2,1/2,0	
SL	3	I(1,1,1)	density	2	add.gen.	1/2,1/2,1/2	
SL	5	C(1,1,1/2)	density	4	add.gen.	0,0,1/2 1/2,1/2,0	
SL	6	I(1,1,1/2)	density	4	add.gen.	1/2,1/2,1/4	
SL	12	P(1/2,1/2,1/2)	density	8	add.gen.	1/2,0,0 0,1/2,0 0,0,1/2	
SL	13	F(1,1,1/2)	density	8	add.gen.	1/2,1/2,0 1/2,0,1/4	
SL	15	I(1/2,1/2,1)	density	8	add.gen.	1/2,0,0 1/4,1/4,1/2	

Wyckoff letter   non-characteristic crystallographic orbits

Wyckoff	SL	coord	group	coord	group
8 e 1	SL 0	x,y,0	P4/mnc (8h)	x,y,1/4	P4$_2$/mbc (8h)
		x,0,z	P4$_2$/nmc (8g)		
	SL 1	x,1/2+x,z	P$\bar{4}2_1$m (4e)	x,1/2+x,1/4	P4/mbm (4g,h)
		x,1/2+x,0	P4/mbm (4g,h)	x,1/2+x,1/8	P4$_2$/mnm (4f,g)
		x,1/2+x,3/8	P4$_2$/mnm (4f,g)		
	SL 2	x,0,3/4	P4$_2$/mcm (4i,j)	x,0,1/4	P4$_2$/mcm (4i,j)
	SL 3	x,x,z	I$\bar{4}$2m (8i)	x,x,0	I4/mmm (8h)
		x,x,1/4	I4/mcm (8h)	x,0,1/2	I4/mmm (8i,j)
		x,0,0	I4/mmm (8i,j)		
	SL 5	1/4,1/4,z	P4/nmm (2c)		
	SL 12	1/4,1/4,0	P4/mmm (1a-d)	1/4,1/4,1/4	P4/mmm (1a-d)
	SL 13	1/4,1/4,3/8	I4/mmm (2a,b)	1/4,1/4,1/8	I4/mmm (2a,b)
	SL 15	1/4,0,3/4	I4/mmm (2a,b)	1/4,0,1/4	I4/mmm (2a,b)
4 d 2	SL 1	0,1/2,z	P4/nmm (2c)		
	SL 5	0,1/2,0	P4/mmm (1a-d)	0,1/2,1/4	P4/mmm (1a-d)
	SL 6	0,1/2,1/8	I4/mmm (2a,b)	0,1/2,3/8	I4/mmm (2a,b)
4 c 2	SL 3	0,0,z	I4/mmm (4e)		
	SL 5	0,0,1/4	P4/mmm (1a-d)		
2 b $\bar{4}$	SL 3	0,0,1/2	I4/mmm (2a,b)		
2 a $\bar{4}$	SL 3	0,0,0	I4/mmm (2a,b)		

Space group No.115 P$\bar{4}$m2

Superlattices

```
SL  0  P(1,1,1)       density 1  add.gen.
SL  1  P(1,1,1/2)     density 2  add.gen.  0,0,1/2
SL  2  C(1,1,1)       density 2  add.gen.  1/2,1/2,0
SL  3  I(1,1,1)       density 2  add.gen.  1/2,1/2,1/2
SL  7  P(1/2,1/2,1)   density 4  add.gen.  1/2,0,0   0,1/2,0
SL 12  P(1/2,1/2,1/2) density 8  add.gen.  1/2,0,0   0,1/2,0   0,0,1/2
```

Wyckoff letter   non-characteristic crystallographic orbits

```
8 l 1    SL 0   x,x,z          P4/mmm (8r)      x,y,1/2      P4/mmm (8p,q)
                x,y,0          P4/mmm (8p,q)    x,1/2+x,z    P4/nmm (8j)
                x,y,1/4        P4₂/mmc (8q)
         SL 1   x,x,1/4        P4/mmm (4j,k)
         SL 2   x,1/2+x,1/2    P4/mmm (4l-o)    x,1/2+x,0    P4/mmm (4l-o)
         SL 3   x,1/2+x,1/4    I4/mmm (8h)
         SL 7   1/4,1/4,z      P4/mmm (2g,h)
         SL12   1/4,1/4,1/4    P4/mmm (1a-d)
4 k m    SL 0   x,1/2,1/2      P4/mmm (4l-o)    x,1/2,0      P4/mmm (4l-o)
                x,1/2,1/4      P4₂/mmc (4j-m)   x,1/2,3/4    P4₂/mmc (4j-m)
4 j m    SL 0   x,0,1/2        P4/mmm (4l-o)    x,0,0        P4/mmm (4l-o)
                x,0,1/4        P4₂/mmc (4j-m)   x,0,3/4      P4₂/mmc (4j-m)
4 i 2    SL 0   x,x,1/2        P4/mmm (4j,k)
         SL 7   1/4,1/4,1/2    P4/mmm (1a-d)
4 h 2    SL 0   x,x,0          P4/mmm (4j,k)
         SL 7   1/4,1/4,0      P4/mmm (1a-d)
2 g mm2  SL 0   0,1/2,z        P4/nmm (2c)
         SL 2   0,1/2,1/2      P4/mmm (1a-d)    0,1/2,0      P4/mmm (1a-d)
         SL 3   0,1/2,1/4      I4/mmm (2a,b)    0,1/2,3/4    I4/mmm (2a,b)
2 f mm2  SL 0   1/2,1/2,z      P4/mmm (2g,h)
         SL 1   1/2,1/2,1/4    P4/mmm (1a-d)
2 e mm2  SL 0   0,0,z          P4/mmm (2g,h)
         SL 1   0,0,1/4        P4/mmm (1a-d)
1 d 4̄m2  SL 0   0,0,1/2        P4/mmm (1a-d)
1 c 4̄m2  SL 0   1/2,1/2,1/2    P4/mmm (1a-d)
1 b 4̄m2  SL 0   1/2,1/2,0      P4/mmm (1a-d)
1 a 4̄m2  SL 0   0,0,0          P4/mmm (1a-d)
```

Space group   No.116   P$\bar{4}$c2

Superlattices

SL	0	P(1,1,1)		density	1	add.gen.			
SL	1	P(1,1,1/2)		density	2	add.gen.	0,0,1/2		
SL	2	C(1,1,1)		density	2	add.gen.	1/2,1/2,0		
SL	3	I(1,1,1)		density	2	add.gen.	1/2,1/2,1/2		
SL	4	P(1,1,1/4)		density	4	add.gen.	0,0,1/4		
SL	5	C(1,1,1/2)		density	4	add.gen.	0,0,1/2	1/2,1/2,0	
SL	6	I(1,1,1/2)		density	4	add.gen.	1/2,1/2,1/4		
SL	8	F(1,1,1)		density	4	add.gen.	0,1/2,1/2	1/2,0,1/2	
SL	12	P(1/2,1/2,1/2)		density	8	add.gen.	1/2,0,0	0,1/2,0	0,0,1/2

Wyckoff letter   non-characteristic crystallographic orbits

Wyckoff letter		SL					
8 j 1		SL 0	x,y,0	P4/mcc (8m)	x,x,z	P4$_2$/mcm (8o)	
			x,y,1/4	P4$_2$/mcm (8n)	x,1/2+x,z	P4$_2$/ncm (8i)	
		SL 1	x,1/2,z	P$\bar{4}$m2 (4j,k)	x,0,z	P$\bar{4}$m2 (4j,k)	
			x,0,0	P4/mmm (4l-o)	x,0,1/4	P4/mmm (4l-o)	
			x,1/2,0	P4/mmm (4l-o)	x,1/2,1/4	P4/mmm (4l-o)	
			x,0,1/8	P4$_2$/mmc (4j-m)	x,0,3/8	P4$_2$/mmc (4j-m)	
			x,1/2,1/8	P4$_2$/mmc (4j-m)	x,1/2,3/8	P4$_2$/mmc (4j-m)	
			x,x,0	P4/mmm (4j,k)			
		SL 2	x,1/2+x,1/4	P4$_2$/mmc (4j-m)	x,1/2+x,3/4	P4$_2$/mmc (4j-m)	
		SL 3	x,1/2+x,0	I4/mcm (8h)			
		SL 8	1/4,1/4,z	I4/mmm (4e)			
		SL12	1/4,1/4,0	P4/mmm (1a-d)			
4 i 2		SL 1	0,1/2,z	P4/nmm (2c)			
		SL 5	0,1/2,1/4	P4/mmm (1a-d)	0,1/2,0	P4/mmm (1a-d)	
		SL 6	0,1/2,1/8	I4/mmm (2a,b)	0,1/2,3/8	I4/mmm (2a,b)	
4 h 2		SL 1	1/2,1/2,z	P4/mmm (2g,h)			
		SL 4	1/2,1/2,1/8	P4/mmm (1a-d)			
4 g 2		SL 1	0,0,z	P4/mmm (2g,h)			
		SL 4	0,0,1/8	P4/mmm (1a-d)			
4 f 2		SL 0	x,x,3/4	P4$_2$/mcm (4i,j)			
		SL 8	1/4,1/4,3/4	I4/mmm (2a,b)			
4 e 2		SL 0	x,x,1/4	P4$_2$/mcm (4i,j)			
		SL 8	1/4,1/4,1/4	I4/mmm (2a,b)			
2 d $\bar{4}$		SL 1	1/2,1/2,0	P4/mmm (1a-d)			
2 c $\bar{4}$		SL 1	0,0,0	P4/mmm (1a-d)			
2 b 222		SL 1	1/2,1/2,1/4	P4/mmm (1a-d)			
2 a 222		SL 1	0,0,1/4	P4/mmm (1a-d)			

Space group No.117 P4̄b2

Superlattices

```
SL  0   P(1,1,1)        density  1   add.gen.
SL  1   P(1,1,1/2)      density  2   add.gen.   0,0,1/2
SL  2   C(1,1,1)        density  2   add.gen.   1/2,1/2,0
SL  3   I(1,1,1)        density  2   add.gen.   1/2,1/2,1/2
SL  5   C(1,1,1/2)      density  4   add.gen.   0,0,1/2   1/2,1/2,0
SL  7   P(1/2,1/2,1)    density  4   add.gen.   1/2,0,0   0,1/2,0
SL 12   P(1/2,1/2,1/2)  density  8   add.gen.   1/2,0,0   0,1/2,0   0,0,1/2
SL 14   C(1/2,1/2,1)    density  8   add.gen.   1/2,0,0   1/4,1/4,0
SL 15   I(1/2,1/2,1)    density  8   add.gen.   1/2,0,0   1/4,1/4,1/2
```

Wyckoff letter   non-characteristic crystallographic orbits

```
  8 i 1     SL 0   x,x,z          P4/nbm (8m)      x,1/2+x,z      P4/mbm (8k)
                   x,y,1/2        P4/mbm (8i,j)    x,y,0          P4/mbm (8i,j)
                   x,y,1/4        P4₂/mbc (8h)
            SL 1   x,1/2+x,1/4    P4/mbm (4g,h)
            SL 2   x,0,z          P4̄2m (4n)        x,0,0          P4/mmm (4j,k)
                   x,0,1/2        P4/mmm (4j,k)    x,0,1/4        P4₂/mcm (4i,j)
                   x,0,3/4        P4₂/mcm (4i,j)   x,x,1/2        P4/mmm (4l-o)
                   x,x,0          P4/mmm (4l-o)
            SL 3   x,x,1/4        I4/mcm (8h)
            SL 7   1/4,0,z        P4/nmm (2c)      1/4,1/4,z      P4/mmm (2g,h)
            SL12   1/4,1/4,1/4    P4/mmm (1a-d)
            SL14   1/4,0,1/2      P4/mmm (1a-d)    1/4,0,0        P4/mmm (1a-d)
            SL15   1/4,0,3/4      I4/mmm (2a,b)    1/4,0,1/4      I4/mmm (2a,b)
  4 h 2     SL 0   x,1/2+x,1/2    P4/mbm (4g,h)
            SL 7   1/4,3/4,1/2    P4/mmm (1a-d)
  4 g 2     SL 0   x,1/2+x,0      P4/mbm (4g,h)
            SL 7   1/4,3/4,0      P4/mmm (1a-d)
  4 f 2     SL 2   0,1/2,z        P4/mmm (2g,h)
            SL 5   0,1/2,1/4      P4/mmm (1a-d)
  4 e 2     SL 2   0,0,z          P4/mmm (2g,h)
            SL 5   0,0,1/4        P4/mmm (1a-d)
  2 d 222   SL 2   0,1/2,1/2      P4/mmm (1a-d)
  2 c 222   SL 2   0,1/2,0        P4/mmm (1a-d)
  2 b 4̄     SL 2   0,0,1/2        P4/mmm (1a-d)
  2 a 4̄     SL 2   0,0,0          P4/mmm (1a-d)
```

Space group No.118 P4̄n2

Superlattices

```
SL   0   P̄(1,1,1)          density  1  add.gen.
SL   1   P(1,1,1/2)        density  2  add.gen.   0,0,1/2
SL   2   C(1,1,1)          density  2  add.gen.   1/2,1/2,0
SL   3   I(1,1,1)          density  2  add.gen.   1/2,1/2,1/2
SL   5   C(1,1,1/2)        density  4  add.gen.   0,0,1/2   1/2,1/2,0
SL   8   F(1,1,1)          density  4  add.gen.   0,1/2,1/2   1/2,0,1/2
SL  12   P(1/2,1/2,1/2)    density  8  add.gen.   1/2,0,0   0,1/2,0   0,0,1/2
```

Wyckoff letter   non-characteristic crystallographic orbits

```
8 i 1      SL 0   x,y,0          P4/mnc (8h)      x,x,z       P4₂/nnm (8m)
                  x,1/2+x,z      P4₂/mnm (8j)     x,y,1/4     P4₂/mnm (8i)
           SL 1   x,1/2+x,0      P4/mbm (4g,h)
           SL 2   x,x,1/4        P4₂/mmc (4j-m)   x,x,3/4     P4₂/mmc (4j-m)
           SL 3   x,0,z          I4̄m2 (8i)        x,0,0       I4/mmm (8i,j)
                  x,0,1/2        I4/mmm (8i,j)    x,0,1/4     I4/mmm (8i,j)
                  x,0,3/4        I4/mmm (8i,j)    1/4,0,1/8   I4₁/amd (8c,d)
                  1/4,0,3/8      I4₁/amd (8c,d)   1/4,0,5/8   I4₁/amd (8c,d)
                  1/4,0,7/8      I4₁/amd (8c,d)   x,x,0       I4/mmm (8h)
           SL 8   1/4,1/4,z      I4/mmm (4e)
           SL12   1/4,1/4,0      P4/mmm (1a-d)
4 h 2      SL 3   0,1/2,z        I4/mmm (4e)
           SL 5   0,1/2,0        P4/mmm (1a-d)
4 g 2      SL 0   x,1/2+x,1/4    P4₂/mnm (4f,g)
           SL 8   1/4,3/4,1/4    I4/mmm (2a,b)
4 f 2      SL 0   x,1/2-x,1/4    P4₂/mnm (4f,g)
           SL 8   1/4,1/4,1/4    I4/mmm (2a,b)
4 e 2      SL 3   0,0,z          I4/mmm (4e)
           SL 5   0,0,1/4        P4/mmm (1a-d)
2 d 222    SL 3   0,1/2,3/4      I4/mmm (2a,b)
2 c 222    SL 3   0,1/2,1/4      I4/mmm (2a,b)
2 b 4̄      SL 3   0,0,1/2        I4/mmm (2a,b)
2 a 4̄      SL 3   0,0,0          I4/mmm (2a,b)
```

Space group   No.119   I4̄m2

Superlattices

```
SL  0   I(1,1,1)        density  1  add.gen.
SL  1   C(1,1,1/2)      density  2  add.gen.  0,0,1/2
SL  3   P(1/2,1/2,1/2)  density  4  add.gen.  1/2,0,0   0,1/2,0
SL  6   P(1/2,1/2,1/4)  density  8  add.gen.  1/2,0,0   0,0,1/4
```

Wyckoff letter   non-characteristic crystallographic orbits

```
16 j 1    SL 0   x,x,z          I4/mmm (16m)    x,1/2+x,z      I4/mmm (16m)
                 x,y,0          I4/mmm (16l)    x,y,1/4        I4/mmm (16l)
                 x,1/4,1/8      I4₁/amd (16f)   x,1/4,3/8      I4₁/amd (16f)
          SL 1   x,x,1/4        P4/mmm (4l-o)   x,1/2+x,0      P4/mmm (4l-o)
          SL 3   1/4,1/4,z      P4/mmm (2g,h)
          SL 6   1/4,1/4,1/8    P4/mmm (1a-d)
 8 i m    SL 0   x,0,0          I4/mmm (8i,j)   x,0,3/4        I4/mmm (8i,j)
                 x,0,1/2        I4/mmm (8i,j)   x,0,1/4        I4/mmm (8i,j)
                 1/4,0,7/8      I4₁/amd (8c,d)  1/4,0,1/8      I4₁/amd (8c,d)
                 1/4,0,3/8      I4₁/amd (8c,d)  1/4,0,5/8      I4₁/amd (8c,d)
 8 h 2    SL 0   x,1/2+x,1/4    I4/mmm (8h)
          SL 3   1/4,3/4,1/4    P4/mmm (1a-d)
 8 g 2    SL 0   x,x,0          I4/mmm (8h)
          SL 3   1/4,1/4,0      P4/mmm (1a-d)
 4 f mm2  SL 0   0,1/2,z        I4/mmm (4e)
          SL 1   0,1/2,0        P4/mmm (1a-d)
 4 e mm2  SL 0   0,0,z          I4/mmm (4e)
          SL 1   0,0,1/4        P4/mmm (1a-d)
 2 d 4̄m2  SL 0   0,1/2,3/4      I4/mmm (2a,b)
 2 c 4̄m2  SL 0   0,1/2,1/4      I4/mmm (2a,b)
 2 b 4̄m2  SL 0   0,0,1/2        I4/mmm (2a,b)
 2 a 4̄m2  SL 0   0,0,0          I4/mmm (2a,b)
```

Space group No.120 I$\bar{4}$c2

Superlattices

SL	0	I(1,1,1)	density	1	add.gen.			
SL	1	C(1,1,1/2)	density	2	add.gen.	0,0,1/2		
SL	2	C(1,1,1/4)	density	4	add.gen.	0,0,1/4		
SL	3	P(1/2,1/2,1/2)	density	4	add.gen.	1/2,0,0	0,1/2,0	
SL	6	P(1/2,1/2,1/4)	density	8	add.gen.	1/2,0,0	0,0,1/4	
SL	8	C(1/2,1/2,1/2)	density	8	add.gen.	1/2,0,0	1/4,1/4,0	
SL	9	I(1/2,1/2,1/2)	density	8	add.gen.	1/2,0,0	0,1/2,0	1/4,1/4,1/4

Wyckoff letter		non-characteristic crystallographic orbits			
16 i 1	SL 0	x,y,0	I4/mcm (16k)	x,y,1/4	I4/mcm (16k)
		x,x,z	I4/mcm (16l)	x,1/2+x,z	I4/mcm (16l)
		x,1/4,1/8	I4$_1$/acd (16e)	x,1/4,3/8	I4$_1$/acd (16e)
	SL 1	x,0,z	P$\bar{4}$2m (4n)	x,0,0	P4/mmm (4j,k)
		x,0,1/4	P4/mmm (4j,k)	x,0,1/8	P4$_2$/mcm (4i,j)
		x,0,3/8	P4$_2$/mcm (4i,j)	x,1/2+x,1/4	P4/mmm (4l-o)
		x,x,0	P4/mmm (4l-o)		
	SL 3	1/4,1/4,z	P4/mmm (2g,h)	1/4,0,z	P4/nmm (2c)
	SL 6	1/4,1/4,1/8	P4/mmm (1a-d)		
	SL 8	1/4,0,0	P4/mmm (1a-d)	1/4,0,1/4	P4/mmm (1a-d)
	SL 9	1/4,0,1/8	I4/mmm (2a,b)	1/4,0,3/8	I4/mmm (2a,b)
8 h 2	SL 0	x,1/2+x,0	I4/mcm (8h)		
	SL 3	1/4,3/4,0	P4/mmm (1a-d)		
8 g 2	SL 1	0,1/2,z	P4/mmm (2g,h)		
	SL 2	0,1/2,1/8	P4/mmm (1a-d)		
8 f 2	SL 1	0,0,z	P4/mmm (2g,h)		
	SL 2	0,0,1/8	P4/mmm (1a-d)		
8 e 2	SL 0	x,x,1/4	I4/mcm (8h)		
	SL 3	1/4,1/4,1/4	P4/mmm (1a-d)		
4 d 222	SL 1	0,1/2,0	P4/mmm (1a-d)		
4 c $\bar{4}$	SL 1	0,1/2,1/4	P4/mmm (1a-d)		
4 b $\bar{4}$	SL 1	0,0,0	P4/mmm (1a-d)		
4 a 222	SL 1	0,0,1/4	P4/mmm (1a-d)		

Space group No.121 I$\bar{4}$2m

Superlattices

```
SL  0  I(1,1,1)         density  1  add.gen.
SL  1  C(1,1,1/2)       density  2  add.gen.  0,0,1/2
SL  2  C(1,1,1/4)       density  4  add.gen.  0,0,1/4
SL  3  P(1/2,1/2,1/2)   density  4  add.gen.  1/2,0,0  0,1/2,0
SL  4  F(1,1,1/2)       density  4  add.gen.  0,1/2,1/4
SL  8  C(1/2,1/2,1/2)   density  8  add.gen.  1/2,0,0  1/4,1/4,0
```

Wyckoff letter   non-characteristic crystallographic orbits

```
16 j 1    SL 0  x,0,z          I4/mmm (16n)   x,y,0        I4/mmm (16l)
                x,y,1/4        I4/mcm (16k)
          SL 1  x,1/2+x,z      P4̄m2 (4j,k)    x,1/2+x,0    P4/mmm (4l-o)
                x,1/2+x,1/4    P4/mmm (4l-o)  x,1/2+x,1/8  P4₂/mmc (4j-m)
                x,1/2+x,3/8    P4₂/mmc (4j-m) x,0,1/4      P4/mmm (4j,k)
          SL 8  1/4,0,1/4      P4/mmm (1a-d)
 8 i m    SL 0  x,x,0          I4/mmm (8h)    x,x,1/4      I4/mcm (8h)
          SL 1  1/4,1/4,z      P4/nmm (2c)
          SL 3  1/4,1/4,0      P4/mmm (1a-d)  1/4,1/4,1/4  P4/mmm (1a-d)
          SL 4  1/4,1/4,3/8    I4/mmm (2a,b)  1/4,1/4,1/8  I4/mmm (2a,b)
 8 h 2    SL 1  0,1/2,z        P4/mmm (2g,h)
          SL 2  0,1/2,1/8      P4/mmm (1a-d)
 8 g 2    SL 0  x,0,1/2        I4/mmm (8i,j)
 8 f 2    SL 0  x,0,0          I4/mmm (8i,j)
 4 e mm2  SL 0  0,0,z          I4/mmm (4e)
          SL 1  0,0,1/4        P4/mmm (1a-d)
 4 d 4̄    SL 1  0,1/2,1/4      P4/mmm (1a-d)
 4 c 222  SL 1  0,1/2,0        P4/mmm (1a-d)
 2 b 4̄2m  SL 0  0,0,1/2        I4/mmm (2a,b)
 2 a 4̄2m  SL 0  0,0,0          I4/mmm (2a,b)
```

Space group No.122 I$\overline{4}$2d

Superlattices

```
SL  0   I(1,1,1)          density  1  add.gen.
SL  1   C(1,1,1/2)        density  2  add.gen.  0,0,1/2
SL  4   F(1,1,1/2)        density  4  add.gen.  0,1/2,1/4
SL  6   P(1/2,1/2,1/4)    density  8  add.gen.  1/2,0,0   0,0,1/4
SL  9   I(1/2,1/2,1/2)    density  8 ,add.gen.  1/2,0,0   0,1/2,0   1/4,1/4,1/4
```

Wyckoff letter   non-characteristic crystallographic orbits

```
16 e 1      SL 0    x,0,z         I4₁/amd (16h)    x,x,0         I4₁/amd (16g)
                    x,x,1/4       I4₁/acd (16f)
            SL 1    x,1/4,3/8     P4₂/mnm (4f,g)
            SL 4    1/4,1/4,z     I4/mmm (4e)
            SL 6    1/4,1/4,0     P4/mmm (1a-d)
            SL 9    1/4,0,1/8     I4/mmm (2a,b)
 8 d 2      SL 0    1/2,1/4,1/8   I4₁/amd (8c,d)   0,1/4,1/8     I4₁/amd (8c,d)
            SL 4    1/4,1/4,1/8   I4/mmm (2a,b)    3/4,1/4,1/8   I4/mmm (2a,b)
 8 c 2      SL 0    0,0,z         I4₁/amd (8e)
            SL 4    0,0,1/4       I4/mmm (2a,b)
 4 b 4̄      SL 0    0,0,1/2       I4₁/amd (4a,b)
 4 a 4̄      SL 0    0,0,0         I4₁/amd (4a,b)
```

Space group   No.123   P4/mmm

Superlattices

```
SL   1   P(1,1,1/2)      density  2  add.gen.  0,0,1/2
SL   2   C(1,1,1)        density  2  add.gen.  1/2,1/2,0
SL   5   C(1,1,1/2)      density  4  add.gen.  0,0,1/2  1/2,1/2,0
SL   7   P(1/2,1/2,1)    density  4  add.gen.  1/2,0,0  0,1/2,0
SL  12   P(1/2,1/2,1/2)  density  8  add.gen.  1/2,0,0  0,1/2,0  0,0,1/2
```

Wyckoff letter   non-characteristic crystallographic orbits

```
 16 u  1      SL  1   x,y,1/4         P4/mmm  (8p,q)
              SL  2   x,1/2+x,z       P4/mmm  (8s,t)
              SL  5   x,1/2+x,1/4     P4/mmm  (4l-o)
  8 t  m      SL  1   x,1/2,1/4       P4/mmm  (4l-o)
  8 s  m      SL  1   x,0,1/4         P4/mmm  (4l-o)
  8 r  m      SL  1   x,x,1/4         P4/mmm  (4j,k)
              SL  7   1/4,1/4,z       P4/mmm  (2g,h)
              SL 12   1/4,1/4,1/4     P4/mmm  (1a-d)
  8 q  m      SL  2   x,1/2+x,1/2     P4/mmm  (4l-o)
  8 p  m      SL  2   x,1/2+x,0       P4/mmm  (4l-o)
  4 k  mm2    SL  7   1/4,1/4,1/2     P4/mmm  (1a-d)
  4 j  mm2    SL  7   1/4,1/4,0       P4/mmm  (1a-d)
  4 i  mm2    SL  2   0,1/2,z         P4/mmm  (2g,h)
              SL  5   0,1/2,1/4       P4/mmm  (1a-d)
  2 h  4mm    SL  1   1/2,1/2,1/4     P4/mmm  (1a-d)
  2 g  4mm    SL  1   0,0,1/4         P4/mmm  (1a-d)
  2 f  mmm    SL  2   0,1/2,0         P4/mmm  (1a-d)
  2 e  mmm    SL  2   0,1/2,1/2       P4/mmm  (1a-d)
```

Space group No.124 P4/mcc

Superlattices

```
SL  1  P(1,1,1/2)        density  2  add.gen.  0,0,1/2
SL  3  I(1,1,1)          density  2  add.gen.  1/2,1/2,1/2
SL  4  P(1,1,1/4)        density  4  add.gen.  0,0,1/4
SL  5  C(1,1,1/2)        density  4  add.gen.  0,0,1/2  1/2,1/2,0
SL 10  C(1,1,1/4)        density  8  add.gen.  0,0,1/4  1/2,1/2,0
SL 12  P(1/2,1/2,1/2)    density  8  add.gen.  1/2,0,0  0,1/2,0  0,0,1/2
SL 19  P(1/2,1/2,1/4)    density 16  add.gen.  1/2,0,0  0,1/2,0  0,0,1/4
```

Wyckoff letter   non-characteristic crystallographic orbits

```
16 n 1     SL 1  x,y,1/4        P4/mmm (8p,q)   x,x,z       P4/mmm (8r)
                 x,1/2,z        P4/mmm (8s,t)   x,0,z       P4/mmm (8s,t)
           SL 3  x,1/2+x,z      I4/mcm (16l)
           SL 4  x,x,1/8        P4/mmm (4j,k)   x,1/2,1/8   P4/mmm (4l-o)
                 x,0,1/8        P4/mmm (4l-o)
           SL 5  x,1/2+x,1/4    P4/mmm (4l-o)
           SL12  1/4,1/4,z      P4/mmm (2g,h)
           SL19  1/4,1/4,1/8    P4/mmm (1a-d)
 8 m m     SL 1  x,x,0          P4/mmm (4j,k)   x,1/2,0     P4/mmm (4l-o)
                 x,0,0          P4/mmm (4l-o)
           SL 3  x,1/2+x,0      I4/mcm (8h)
           SL12  1/4,1/4,0      P4/mmm (1a-d)
 8 l 2     SL 1  x,1/2,1/4      P4/mmm (4l-o)
 8 k 2     SL 1  x,0,1/4        P4/mmm (4l-o)
 8 j 2     SL 1  x,x,1/4        P4/mmm (4j,k)
           SL12  1/4,1/4,1/4    P4/mmm (1a-d)
 8 i 2     SL 5  0,1/2,z        P4/mmm (2g,h)
           SL10  0,1/2,1/8      P4/mmm (1a-d)
 4 h 4     SL 1  1/2,1/2,z      P4/mmm (2g,h)
           SL 4  1/2,1/2,1/8    P4/mmm (1a-d)
 4 g 4     SL 1  0,0,z          P4/mmm (2g,h)
           SL 4  0,0,1/8        P4/mmm (1a-d)
 4 f 222   SL 5  0,1/2,1/4      P4/mmm (1a-d)
 4 e 2/m   SL 5  0,1/2,0        P4/mmm (1a-d)
 2 d 4/m   SL 1  1/2,1/2,0      P4/mmm (1a-d)
 2 c 422   SL 1  1/2,1/2,1/4    P4/mmm (1a-d)
 2 b 4/m   SL 1  0,0,0          P4/mmm (1a-d)
 2 a 422   SL 1  0,0,1/4        P4/mmm (1a-d)
```

Space group No.125 P4/nbm

Superlattices

```
SL  2  C(1,1,1)        density  2  add.gen.  1/2,1/2,0
SL  3  I(1,1,1)        density  2  add.gen.  1/2,1/2,1/2
SL  5  C(1,1,1/2)      density  4  add.gen.  0,0,1/2   1/2,1/2,0
SL  7  P(1/2,1/2,1)    density  4  add.gen.  1/2,0,0   0,1/2,0
SL 12  P(1/2,1/2,1/2)  density  8  add.gen.  1/2,0,0   0,1/2,0   0,0,1/2
SL 14  C(1/2,1/2,1)    density  8  add.gen.  1/2,0,0   1/4,1/4,0
SL 21  C(1/2,1/2,1/2)  density 16  add.gen.  1/2,0,0   1/4,1/4,0   0,0,1/2
```

Wyckoff letter   non-characteristic crystallographic orbits

```
16 n 1      SL  2   x,1/4,z      P4/mmm (8r)    x,x,z       P4/mmm (8s,t)
                    x,y,1/2      P4/mmm (8p,q)  x,y,0       P4/mmm (8p,q)
            SL  3   x,y,1/4      I4/mcm (16k)
            SL  5   x,x,1/4      P4/mmm (4l-o)  x,1/4,1/4   P4/mmm (4j,k)
            SL  7   x,0,1/2      P4/mmm (4l-o)  x,0,0       P4/mmm (4l-o)
            SL 14   1/4,0,z      P4/mmm (2g,h)
            SL 21   1/4,0,1/4    P4/mmm (1a-d)
 8 m m      SL  2   x,-x,0       P4/mmm (4l-o)  x,-x,1/2    P4/mmm (4l-o)
            SL  3   x,-x,1/4     I4/mcm (8h)
            SL  7   0,0,z        P4/mmm (2g,h)
            SL 12   0,0,1/4      P4/mmm (1a-d)
 8 l 2      SL  2   x,1/4,1/2    P4/mmm (4j,k)
            SL 14   0,1/4,1/2    P4/mmm (1a-d)
 8 k 2      SL  2   x,1/4,0      P4/mmm (4j,k)
            SL 14   0,1/4,0      P4/mmm (1a-d)
 8 j 2      SL  2   x,x,1/2      P4/mmm (4l-o)
 8 i 2      SL  2   x,x,0        P4/mmm (4l-o)
 4 h mm2    SL  2   3/4,1/4,z    P4/mmm (2g,h)
            SL  5   3/4,1/4,1/4  P4/mmm (1a-d)
 4 g 4      SL  2   1/4,1/4,z    P4/mmm (2g,h)
            SL  5   1/4,1/4,1/4  P4/mmm (1a-d)
 4 f 2/m    SL  7   0,0,1/2      P4/mmm (1a-d)
 4 e 2/m    SL  7   0,0,0        P4/mmm (1a-d)
 2 d 42m    SL  2   3/4,1/4,1/2  P4/mmm (1a-d)
 2 c 42m    SL  2   3/4,1/4,0    P4/mmm (1a-d)
 2 b 422    SL  2   1/4,1/4,1/2  P4/mmm (1a-d)
 2 a 422    SL  2   1/4,1/4,0    P4/mmm (1a-d)
```

Space group No.126 P4/nnc

Superlattices

```
SL  1   P(1,1,1/2)      density  2   add.gen.  0,0,1/2
SL  2   C(1,1,1)        density  2   add.gen.  1/2,1/2,0
SL  3   I(1,1,1)        density  2   add.gen.  1/2,1/2,1/2
SL  5   C(1,1,1/2)      density  4   add.gen.  0,0,1/2   1/2,1/2,0
SL  6   I(1,1,1/2)      density  4   add.gen.  1/2,1/2,1/4
SL  8   F(1,1,1)        density  4   add.gen.  0,1/2,1/2  1/2,0,1/2
SL 10   C(1,1,1/4)      density  8   add.gen.  0,0,1/4   1/2,1/2,0
SL 12   P(1/2,1/2,1/2)  density  8   add.gen.  1/2,0,0   0,1/2,0   0,0,1/2
SL 19   P(1/2,1/2,1/4)  density 16   add.gen.  1/2,0,0   0,1/2,0   0,0,1/4
SL 21   C(1/2,1/2,1/2)  density 16   add.gen.  1/2,0,0   1/4,1/4,0   0,0,1/2
```

Wyckoff letter   non-characteristic crystallographic orbits

```
16 k 1      SL  1   x,1/2+x,z      P4/nbm (8m)
            SL  2   x,y,0          P4/mcc (8m)
            SL  3   x,y,1/4        I4/mmm (16l)     x,x,z         I4/mmm (16m)
                    x,1/4,z        I4/mmm (16n)
            SL  5   x,1/2+x,0      P4/mmm (4l-o)    x,1/2+x,1/4   P4/mmm (4l-o)
                    x,x,0          P4/mmm (4l-o)    x,1/4,0       P4/mmm (4j,k)
            SL  6   x,1/2+x,1/8    I4/mcm (8h)
            SL  8   x,0,0          I4/mcm (8h)
            SL 12   0,0,z          P4/mmm (2g,h)
            SL 19   0,0,1/8        P4/mmm (1a-d)
            SL 21   1/4,0,0        P4/mmm (1a-d)
 8 j 2      SL  3   x,3/4,1/4      I4/mmm (8i,j)
 8 i 2      SL  3   x,1/4,1/4      I4/mmm (8i,j)
 8 h 2      SL  3   x,x,1/4        I4/mmm (8h)
            SL 12   0,0,1/4        P4/mmm (1a-d)
 8 g 2      SL  5   1/4,3/4,z      P4/mmm (2g,h)
            SL 10   1/4,3/4,1/8    P4/mmm (1a-d)
 8 f 1̄      SL 12   0,0,0          P4/mmm (1a-d)
 4 e 4      SL  3   1/4,1/4,z      I4/mmm (4e)
            SL  5   1/4,1/4,0      P4/mmm (1a-d)
 4 d 4̄      SL  5   1/4,3/4,0      P4/mmm (1a-d)
 4 c 222    SL  5   1/4,3/4,3/4    P4/mmm (1a-d)
 2 b 422    SL  3   1/4,1/4,3/4    I4/mmm (2a,b)
 2 a 422    SL  3   1/4,1/4,1/4    I4/mmm (2a,b)
```

Space group No.127 P4/mbm

Superlattices

```
SL  1  P(1,1,1/2)      density  2  add.gen.  0,0,1/2
SL  2  C(1,1,1)        density  2  add.gen.  1/2,1/2,0
SL  5  C(1,1,1/2)      density  4  add.gen.  0,0,1/2   1/2,1/2,0
SL  7  P(1/2,1/2,1)    density  4  add.gen.  1/2,0,0   0,1/2,0
SL 12  P(1/2,1/2,1/2)  density  8  add.gen.  1/2,0,0   0,1/2,0   0,0,1/2
SL 14  C(1/2,1/2,1)    density  8  add.gen.  1/2,0,0   1/4,1/4,0
SL 21  C(1/2,1/2,1/2)  density 16  add.gen.  1/2,0,0   1/4,1/4,0   0,0,1/2
```

Wyckoff letter   non-characteristic crystallographic orbits

```
  16 l l    SL  1  x,y,1/4        P4/mbm (8i,j)
            SL  2  x,x,z          P4/mmm (8s,t)   x,0,z     P4/mmm (8r)
            SL  5  x,x,1/4        P4/mmm (4l-o)   x,0,1/4   P4/mmm (4j,k)
            SL14   1/4,0,z        P4/mmm (2g,h)
            SL21   1/4,0,1/4      P4/mmm (1a-d)
   8 k m    SL  1  x,1/2+x,1/4    P4/mbm (4g,h)
            SL  7  1/4,3/4,z      P4/mmm (2g,h)
            SL12   1/4,3/4,1/4    P4/mmm (1a-d)
   8 j m    SL  2  x,x,1/2        P4/mmm (4l-o)   x,0,1/2   P4/mmm (4j,k)
            SL14   1/4,0,1/2      P4/mmm (1a-d)
   8 i m    SL  2  x,x,0          P4/mmm (4l-o)   x,0,0     P4/mmm (4j,k)
            SL14   1/4,0,0        P4/mmm (1a-d)
   4 h mm2  SL  7  1/4,3/4,1/2    P4/mmm (1a-d)
   4 g mm2  SL  7  1/4,3/4,0      P4/mmm (1a-d)
   4 f mm2  SL  2  0,1/2,z        P4/mmm (2g,h)
            SL  5  0,1/2,1/4      P4/mmm (1a-d)
   4 e 4    SL  2  0,0,z          P4/mmm (2g,h)
            SL  5  0,0,1/4        P4/mmm (1a-d)
   2 d mmm  SL  2  0,1/2,0        P4/mmm (1a-d)
   2 c mmm  SL  2  0,1/2,1/2      P4/mmm (1a-d)
   2 b 4/m  SL  2  0,0,1/2        P4/mmm (1a-d)
   2 a 4/m  SL  2  0,0,0          P4/mmm (1a-d)
```

Space group   No.128   P4/mnc

Superlattices

```
SL  1  P(1,1,1/2)        density  2  add.gen.  0,0,1/2
SL  3  I(1,1,1)          density  2  add.gen.  1/2,1/2,1/2
SL  4  P(1,1,1/4)        density  4  add.gen.  0,0,1/4
SL  5  C(1,1,1/2)        density  4  add.gen.  0,0,1/2   1/2,1/2,0
SL 10  C(1,1,1/4)        density  8  add.gen.  0,0,1/4   1/2,1/2,0
SL 12  P(1/2,1/2,1/2)    density  8  add.gen.  1/2,0,0   0,1/2,0   0,0,1/2
SL 19  P(1/2,1/2,1/4)    density 16  add.gen.  1/2,0,0   0,1/2,0   0,0,1/4
SL 21  C(1/2,1/2,1/2)    density 16  add.gen.  1/2,0,0   1/4,1/4,0   0,0,1/2
```

Wyckoff letter   non-characteristic crystallographic orbits

```
16 i 1      SL  1  x,y,1/4        P4/mbm (8i,j)    x,1/2+x,z    P4/mbm (8k)
            SL  3  x,x,z          I4/mmm (16m)     x,0,z        I4/mmm (16n)
            SL  4  x,1/2+x,1/8    P4/mbm (4g,h)
            SL  5  x,x,1/4        P4/mmm (4l-o)    x,0,1/4      P4/mmm (4j,k)
            SL12   1/4,1/4,z      P4/mmm (2g,h)
            SL19   1/4,1/4,1/8    P4/mmm (1a-d)
            SL21   1/4,0,1/4      P4/mmm (1a-d)
 8 h m      SL  1  x,1/2+x,0      P4/mbm (4g,h)
            SL  3  x,x,0          I4/mmm (8h)      x,1/2,0      I4/mmm (8i,j)
                   x,0,0          I4/mmm (8i,j)
            SL12   1/4,1/4,0      P4/mmm (1a-d)
 8 g 2      SL  1  x,1/2+x,1/4    P4/mbm (4g,h)
            SL12   1/4,3/4,1/4    P4/mmm (1a-d)
 8 f 2      SL  5  0,1/2,z        P4/mmm (2g,h)
            SL10   0,1/2,1/8      P4/mmm (1a-d)
 4 e 4      SL  3  0,0,z          I4/mmm (4e)
            SL  5  0,0,1/4        P4/mmm (1a-d)
 4 d 222    SL  5  0,1/2,1/4      P4/mmm (1a-d)
 4 c 2/m    SL  5  0,1/2,0        P4/mmm (1a-d)
 2 b 4/m    SL  3  0,0,1/2        I4/mmm (2a,b)
 2 a 4/m    SL  3  0,0,0          I4/mmm (2a,b)
```

Space group   No.129   P4/nmm

Superlattices

SL	2	C(1,1,1)	density 2	add.gen.	1/2,1/2,0		
SL	3	I(1,1,1)	density 2	add.gen.	1/2,1/2,1/2		
SL	5	C(1,1,1/2)	density 4	add.gen.	0,0,1/2	1/2,1/2,0	
SL	7	P(1/2,1/2,1)	density 4	add.gen.	1/2,0,0	0,1/2,0	
SL	12	P(1/2,1/2,1/2)	density 8	add.gen.	1/2,0,0	0,1/2,0	0,0,1/2
SL	14	C(1/2,1/2,1)	density 8	add.gen.	1/2,0,0	1/4,1/4,0	

Wyckoff letter   non-characteristic crystallographic orbits

Wyckoff			SL	orbit	group(pos)	orbit2	group2(pos2)
16	k	1	SL 2	x,1/2+x,z	P4/mmm (8s,t)	x,y,1/2	P4/mmm (8p,q)
				x,y,0	P4/mmm (8p,q)		
			SL 3	x,y,1/4	I4/mmm (16l)		
			SL 5	x,1/2+x,1/4	P4/mmm (4l-o)		
			SL 7	x,0,1/2	P4/mmm (4l-o)	x,0,0	P4/mmm (4l-o)
8	j	m	SL 2	x,x,1/2	P4/mmm (4l-o)	x,x,0	P4/mmm (4l-o)
			SL 3	x,x,1/4	I4/mmm (8h)		
			SL 7	0,0,z	P4/mmm (2g,h)		
			SL12	0,0,1/4	P4/mmm (1a-d)		
8	i	m	SL 2	1/4,y,1/2	P4/mmm (4j,k)	1/4,y,0	P4/mmm (4j,k)
			SL 3	1/4,y,1/4	I4/mmm (8i,j)	1/4,y,3/4	I4/mmm (8i,j)
			SL14	1/4,0,1/2	P4/mmm (1a-d)	1/4,0,0	P4/mmm (1a-d)
8	h	2	SL 2	x,-x,1/2	P4/mmm (4l-o)		
8	g	2	SL 2	x,-x,0	P4/mmm (4l-o)		
4	f	mm2	SL 2	3/4,1/4,z	P4/mmm (2g,h)		
			SL 5	3/4,1/4,1/4	P4/mmm (1a-d)		
4	e	2/m	SL 7	0,0,1/2	P4/mmm (1a-d)		
4	d	2/m	SL 7	0,0,0	P4/mmm (1a-d)		
2	c	4mm	SL 2	1/4,1/4,1/2	P4/mmm (1a-d)	1/4,1/4,0	P4/mmm (1a-d)
			SL 3	1/4,1/4,1/4	I4/mmm (2a,b)	1/4,1/4,3/4	I4/mmm (2a,b)
2	b	4̄m2	SL 2	3/4,1/4,1/2	P4/mmm (1a-d)		
2	a	4̄m2	SL 2	3/4,1/4,0	P4/mmm (1a-d)		

Space group   No.130   P4/ncc

Superlattices

```
SL  1   P(1,1,1/2)      density  2   add.gen.   0,0,1/2
SL  2   C(1,1,1)        density  2   add.gen.   1/2,1/2,0
SL  3   I(1,1,1)        density  2   add.gen.   1/2,1/2,1/2
SL  5   C(1,1,1/2)      density  4   add.gen.   0,0,1/2   1/2,1/2,0
SL  6   I(1,1,1/2)      density  4   add.gen.   1/2,1/2,1/4
SL  8   F(1,1,1)        density  4   add.gen.   0,1/2,1/2   1/2,0,1/2
SL 10   C(1,1,1/4)      density  8   add.gen.   0,0,1/4   1/2,1/2,0
SL 12   P(1/2,1/2,1/2)  density  8   add.gen.   1/2,0,0   0,1/2,0   0,0,1/2
SL 19   P(1/2,1/2,1/4)  density 16   add.gen.   1/2,0,0   0,1/2,0   0,0,1/4
SL 21   C(1/2,1/2,1/2)  density 16   add.gen.   1/2,0,0   1/4,1/4,0   0,0,1/2
```

Wyckoff letter   non-characteristic crystallographic orbits

16 g 1	SL 1	x,1/4,z	P4/nmm (8i)	x,x,z	P4/nmm (8j)
	SL 2	x,y,0	P4/mcc (8m)		
	SL 3	x,y,1/4	I4/mcm (16k)	x,1/2+x,z	I4/mcm (16l)
	SL 5	x,x,0	P4/mmm (4l-o)	x,x,1/4	P4/mmm (4l-o)
		x,1/4,1/4	P4/mmm (4j,k)	x,1/4,0	P4/mmm (4j,k)
		x,1/2+x,0	P4/mmm (4l-o)		
	SL 6	x,x,1/8	I4/mmm (8h)	x,1/4,1/8	I4/mmm (8i,j)
		x,3/4,1/8	I4/mmm (8i,j)		
	SL 8	x,0,0	I4/mcm (8h)		
	SL12	0,0,z	P4/mmm (2g,h)		
	SL19	0,0,1/8	P4/mmm (1a-d)		
	SL21	1/4,0,0	P4/mmm (1a-d)	1/4,0,1/4	P4/mmm (1a-d)
8 f 2	SL 3	x,-x,1/4	I4/mcm (8h)		
	SL12	0,0,1/4	P4/mmm (1a-d)		
8 e 2	SL 5	3/4,1/4,z	P4/mmm (2g,h)		
	SL10	3/4,1/4,1/8	P4/mmm (1a-d)		
8 d 1̄	SL12	0,0,0	P4/mmm (1a-d)		
4 c 4	SL 1	1/4,1/4,z	P4/nmm (2c)		
	SL 5	1/4,1/4,1/4	P4/mmm (1a-d)	1/4,1/4,0	P4/mmm (1a-d)
	SL 6	1/4,1/4,1/8	I4/mmm (2a,b)	1/4,1/4,3/8	I4/mmm (2a,b)
4 b 4̄	SL 5	3/4,1/4,0	P4/mmm (1a-d)		
4 a 222	SL 5	3/4,1/4,1/4	P4/mmm (1a-d)		

Space group   No.131   P4$_2$/mmc

Superlattices

```
SL  1  P(1,1,1/2)      density  2  add.gen.  0,0,1/2
SL  3  I(1,1,1)        density  2  add.gen.  1/2,1/2,1/2
SL  4  P(1,1,1/4)      density  4  add.gen.  0,0,1/4
SL  5  C(1,1,1/2)      density  4  add.gen.  0,0,1/2   1/2,1/2,0
SL 12  P(1/2,1/2,1/2)  density  8  add.gen.  1/2,0,0  0,1/2,0  0,0,1/2
SL 19  P(1/2,1/2,1/4)  density 16  add.gen.  1/2,0,0  0,1/2,0  0,0,1/4
```

Wyckoff letter   non-characteristic crystallographic orbits

```
 16 r 1      SL 1   x,y,1/4         P4/mmm (8p,q)   x,x,z      P4/mmm (8r)
             SL 3   x,1/2+x,z       I4/mmm (16m)
             SL 4   x,x,1/8         P4/mmm (4j,k)
             SL 5   x,1/2+x,1/4     P4/mmm (4l-o)
             SL12   1/4,1/4,z       P4/mmm (2g,h)
             SL19   1/4,1/4,1/8     P4/mmm (1a-d)
  8 q m      SL 1   x,x,0           P4/mmm (4j,k)
             SL 3   x,1/2+x,0       I4/mmm (8h)
             SL12   1/4,1/4,0       P4/mmm (1a-d)
  8 p m      SL 1   1/2,y,1/4       P4/mmm (4l-o)
  8 o m      SL 1   0,y,1/4         P4/mmm (4l-o)
  8 n 2      SL 1   x,x,1/4         P4/mmm (4j,k)
             SL12   1/4,1/4,1/4     P4/mmm (1a-d)
  4 l mm2    SL 3   0,1/2,z         I4/mmm (4e)
             SL 5   0,1/2,1/4       P4/mmm (1a-d)
  4 h mm2    SL 1   1/2,1/2,z       P4/mmm (2g,h)
             SL 4   1/2,1/2,1/8     P4/mmm (1a-d)
  4 g mm2    SL 1   0,0,z           P4/mmm (2g,h)
             SL 4   0,0,1/8         P4/mmm (1a-d)
  2 f 4̄m2    SL 1   1/2,1/2,1/4     P4/mmm (1a-d)
  2 e 4̄m2    SL 1   0,0,1/4         P4/mmm (1a-d)
  2 d mmm    SL 3   0,1/2,1/2       I4/mmm (2a,b)
  2 c mmm    SL 3   0,1/2,0         I4/mmm (2a,b)
  2 b mmm    SL 1   1/2,1/2,0       P4/mmm (1a-d)
  2 a mmm    SL 1   0,0,0           P4/mmm (1a-d)
```

Space group   No.132   P4$_2$/mcm

Superlattices

```
SL  1  P(1,1,1/2)        density 2  add.gen.  0,0,1/2
SL  2  C(1,1,1)          density 2  add.gen.  1/2,1/2,0
SL  4  P(1,1,1/4)        density 4  add.gen.  0,0,1/4
SL  5  C(1,1,1/2)        density 4  add.gen.  0,0,1/2  1/2,1/2,0
SL  8  F(1,1,1)          density 4  add.gen.  0,1/2,1/2  1/2,0,1/2
SL 10  C(1,1,1/4)        density 8  add.gen.  0,0,1/4  1/2,1/2,0
SL 12  P(1/2,1/2,1/2)    density 8  add.gen.  1/2,0,0  0,1/2,0  0,0,1/2
```

Wyckoff letter   non-characteristic crystallographic orbits

```
16 p  1     SL  1  x,y,1/4        P4/mmm (8p,q)   x,1/2,z    P4/mmm (8s,t).
                   x,0,z          P4/mmm (8s,t)
            SL  2  x,1/2+x,z      P4₂/mmc (8o,p)
            SL  4  x,1/2,1/8      P4/mmm (4l-o)   x,0,1/8    P4/mmm (4l-o)
            SL  5  x,1/2+x,1/4    P4/mmm (4l-o)
 8 o  m     SL  1  x,x,1/4        P4/mmm (4j,k)
            SL  8  1/4,1/4,z      I4/mmm (4e)
            SL 12  1/4,1/4,1/4    P4/mmm (1a-d)
 8 n  m     SL  1  x,1/2,0        P4/mmm (4l-o)   x,0,0      P4/mmm (4l-o)
            SL  2  x,1/2+x,0      P4₂/mmc (4j-m)  x,1/2-x,0  P4₂/mmc (4j-m)
 8 m  2     SL  1  x,1/2,1/4      P4/mmm (4l-o)
 8 l  2     SL  1  x,0,1/4        P4/mmm (4l-o)
 8 k  2     SL  5  0,1/2,z        P4/mmm (2g,h)
            SL 10  0,1/2,1/8      P4/mmm (1a-d)
 4 j  mm2   SL  8  1/4,1/4,1/2    I4/mmm (2a,b)
 4 i  mm2   SL  8  1/4,1/4,0      I4/mmm (2a,b)
 4 h  mm2   SL  1  1/2,1/2,z      P4/mmm (2g,h)
            SL  4  1/2,1/2,1/8    P4/mmm (1a-d)
 4 g  mm2   SL  1  0,0,z          P4/mmm (2g,h)
            SL  4  0,0,1/8        P4/mmm (1a-d)
 4 f  2/m   SL  5  0,1/2,0        P4/mmm (1a-d)
 4 e  222   SL  5  0,1/2,1/4      P4/mmm (1a-d)
 2 d  4̄2m   SL  1  1/2,1/2,1/4    P4/mmm (1a-d)
 2 c  mmm   SL  1  1/2,1/2,0      P4/mmm (1a-d)
 2 b  4̄2m   SL  1  0,0,1/4        P4/mmm (1a-d)
 2 a  mmm   SL  1  0,0,0          P4/mmm (1a-d)
```

Space group No.133 P4$_2$/nbc

Superlattices

```
SL  1  P(1,1,1/2)      density  2  add.gen.  0,0,1/2
SL  2  C(1,1,1)        density  2  add.gen.  1/2,1/2,0
SL  3  I(1,1,1)        density  2  add.gen.  1/2,1/2,1/2
SL  5  C(1,1,1/2)      density  4  add.gen.  0,0,1/2  1/2,1/2,0
SL  6  I(1,1,1/2)      density  4  add.gen.  1/2,1/2,1/4
SL  7  P(1/2,1/2,1)    density  4  add.gen.  1/2,0,0  0,1/2,0
SL 10  C(1,1,1/4)      density  8  add.gen.  0,0,1/4  1/2,1/2,0
SL 12  P(1/2,1/2,1/2)  density  8  add.gen.  1/2,0,0  0,1/2,0  0,0,1/2
SL 15  I(1/2,1/2,1)    density  8  add.gen.  1/2,0,0  1/4,1/4,1/2
SL 19  P(1/2,1/2,1/4)  density 16  add.gen.  1/2,0,0  0,1/2,0  0,0,1/4
SL 21  C(1/2,1/2,1/2)  density 16  add.gen.  1/2,0,0  1/4,1/4,0  0,0,1/2
```

Wyckoff letter   non-characteristic crystallographic orbits

```
16 k 1      SL 1   x,1/2+x,z     P4/nbm (8m)
            SL 2   x,1/4,z       P4₂/mcm (8o)      x,y,0         P4₂/mcm (8n)
            SL 3   x,y,1/4       I4/mcm (16k)      x,x,z         I4/mcm (16l)
            SL 5   x,1/4,1/4     P4/mmm (4j,k)     x,1/2+x,0     P4/mmm (4l-o)
                   x,1/2+x,1/4   P4/mmm (4l-o)     x,x,0         P4/mmm (4l-o)
            SL 6   x,1/2+x,1/8   I4/mcm (8h)
            SL 7   x,0,0         P4₂/mmc (4j-m)    x,0,1/2       P4₂/mmc (4j-m)
            SL12   0,0,z         P4/mmm (2g,h)
            SL15   1/4,0,z       I4/mmm (4e)
            SL19   0,0,1/8       P4/mmm (1a-d)
            SL21   1/4,0,1/4     P4/mmm (1a-d)
 8 j 2      SL 3   x,x,1/4       I4/mcm (8h)
            SL12   0,0,1/4       P4/mmm (1a-d)
 8 i 2      SL 2   x,1/4,1/2     P4₂/mcm (4i,j)
            SL15   0,1/4,1/2     I4/mmm (2a,b)
 8 h 2      SL 2   x,1/4,0       P4₂/mcm (4i,j)
            SL15   0,1/4,0       I4/mmm (2a,b)
 8 g 2      SL 5   3/4,1/4,z     P4/mmm (2g,h)
            SL10   3/4,1/4,1/8   P4/mmm (1a-d)
 8 f 2      SL 5   1/4,1/4,z     P4/mmm (2g,h)
            SL10   1/4,1/4,1/8   P4/mmm (1a-d)
 8 e 1̄      SL12   0,0,0         P4/mmm (1a-d)
 4 d 4̄      SL 5   3/4,1/4,3/4   P4/mmm (1a-d)
 4 c 222    SL 5   1/4,1/4,1/4   P4/mmm (1a-d)
 4 b 222    SL 5   3/4,1/4,0     P4/mmm (1a-d)
 4 a 222    SL 5   1/4,1/4,0     P4/mmm (1a-d)
```

Space group   No.134   P4$_2$/nnm

Superlattices

SL	2	C(1,1,1)	density	2	add.gen.	1/2,1/2,0		
SL	3	I(1,1,1)	density	2	add.gen.	1/2,1/2,1/2		
SL	5	C(1,1,1/2)	density	4	add.gen.	0,0,1/2	1/2,1/2,0	
SL	8	F(1,1,1)	density	4	add.gen.	0,1/2,1/2	1/2,0,1/2	
SL	10	C(1,1,1/4)	density	8	add.gen.	0,0,1/4	1/2,1/2,0	
SL	12	P(1/2,1/2,1/2)	density	8	add.gen.	1/2,0,0	0,1/2,0	0,0,1/2
SL	21	C(1/2,1/2,1/2)	density	16	add.gen.	1/2,0,0	1/4,1/4,0	0,0,1/2

Wyckoff letter   non-characteristic crystallographic orbits

16 n 1	SL 2	x,x,z	P4$_2$/mmc (8o,p)	x,y,0	P4$_2$/mmc (8q)
	SL 3	x,y,1/4	I4/mmm (16l)	x,1/4,z	I4/mmm (16n)
	SL 5	x,x,1/4	P4/mmm (4l-o)	x,1/4,0	P4/mmm (4j,k)
	SL 8	x,0,0	I4/mmm (8h)		
	SL21	1/4,0,0	P4/mmm (1a-d)		
8 m m	SL 2	x,-x,0	P4$_2$/mmc (4j-m)	x,-x,1/2	P4$_2$/mmc (4j-m)
	SL 3	x,-x,1/4	I4/mmm (8h)		
	SL 8	0,0,z	I4/mmm (4e)		
	SL12	0,0,1/4	P4/mmm (1a-d)		
8 l 2	SL 2	x,x,1/2	P4$_2$/mmc (4j-m)		
8 k 2	SL 2	x,x,0	P4$_2$/mmc (4j-m)		
8 j 2	SL 3	x,1/4,1/4	I4/mmm (8i,j)		
8 i 2	SL 3	x,1/4,3/4	I4/mmm (8i,j)		
8 h 2	SL 5	1/4,1/4,z	P4/mmm (2g,h)		
	SL10	1/4,1/4,1/8	P4/mmm (1a-d)		
4 g mm2	SL 3	3/4,1/4,z	I4/mmm (4e)		
	SL 5	3/4,1/4,0	P4/mmm (1a-d)		
4 f 2/m	SL 8	0,0,0	I4/mmm (2a,b)		
4 e 2/m	SL 8	0,0,1/2	I4/mmm (2a,b)		
4 d 222	SL 5	1/4,1/4,0	P4/mmm (1a-d)		
4 c 222	SL 5	1/4,1/4,1/4	P4/mmm (1a-d)		
2 b $\bar{4}$2m	SL 3	3/4,1/4,1/4	I4/mmm (2a,b)		
2 a $\bar{4}$2m	SL 3	1/4,3/4,1/4	I4/mmm (2a,b)		

Space group No.135 P4$_2$/mbc

Superlattices

```
SL  1  P(1,1,1/2)       density  2  add.gen.  0,0,1/2
SL  2  C(1,1,1)         density  2  add.gen.  1/2,1/2,0
SL  3  I(1,1,1)         density  2  add.gen.  1/2,1/2,1/2
SL  4  P(1,1,1/4)       density  4  add.gen.  0,0,1/4
SL  5  C(1,1,1/2)       density  4  add.gen.  0,0,1/2   1/2,1/2,0
SL 10  C(1,1,1/4)       density  8  add.gen.  0,0,1/4   1/2,1/2,0
SL 12  P(1/2,1/2,1/2)   density  8  add.gen.  1/2,0,0   0,1/2,0   0,0,1/2
SL 15  I(1/2,1/2,1)     density  8  add.gen.  1/2,0,0   1/4,1/4,1/2
SL 19  P(1/2,1/2,1/4)   density 16  add.gen.  1/2,0,0   0,1/2,0   0,0,1/4
SL 21  C(1/2,1/2,1/2)   density 16  add.gen.  1/2,0,0   1/4,1/4,0   0,0,1/2
```

Wyckoff letter   non-characteristic crystallographic orbits

```
16 i 1     SL 1  x,y,1/4        P4/mbm (8i,j)   x,1/2+x,z    P4/mbm (8k)
           SL 2  x,0,z          P4₂/mcm (8o)
           SL 3  x,x,z          I4/mcm (16l)
           SL 4  x,1/2+x,1/8    P4/mbm (4g,h)
           SL 5  x,0,1/4        P4/mmm (4j,k)    x,x,1/4      P4/mmm (4l-o)
           SL12  1/4,1/4,z      P4/mmm (2g,h)
           SL15  1/4,0,z        I4/mmm (4e)
           SL19  1/4,1/4,1/8    P4/mmm (1a-d)
           SL21  1/4,0,1/4      P4/mmm (1a-d)
 8 h m     SL 1  x,1/2+x,0      P4/mbm (4g,h)
           SL 2  x,0,0          P4₂/mcm (4i,j)   0,y,0        P4₂/mcm (4i,j)
           SL 3  x,x,0          I4/mcm (8h)
           SL12  1/4,1/4,0      P4/mmm (1a-d)
           SL15  1/4,0,0        I4/mmm (2a,b)    0,1/4,0      I4/mmm (2a,b)
 8 g 2     SL 1  x,1/2+x,1/4    P4/mbm (4g,h)
           SL12  1/4,3/4,1/4    P4/mmm (1a-d)
 8 f 2     SL 5  0,1/2,z        P4/mmm (2g,h)
           SL10  0,1/2,1/8      P4/mmm (1a-d)
 8 e 2     SL 5  0,0,z          P4/mmm (2g,h)
           SL10  0,0,1/8        P4/mmm (1a-d)
 4 d 222   SL 5  0,1/2,1/4      P4/mmm (1a-d)
 4 c 2/m   SL 5  0,1/2,0        P4/mmm (1a-d)
 4 b 4̄     SL 5  0,0,1/4        P4/mmm (1a-d)
 4 a 2/m   SL 5  0,0,0          P4/mmm (1a-d)
```

Space group   No.136   P4₂/mnm

Superlattices

```
SL  1  P(1,1,1/2)       density  2  add.gen.  0,0,1/2
SL  2  C(1,1,1)         density  2  add.gen.  1/2,1/2,0
SL  3  I(1,1,1)         density  2  add.gen.  1/2,1/2,1/2
SL  5  C(1,1,1/2)       density  4  add.gen.  0,0,1/2  1/2,1/2,0
SL  8  F(1,1,1)         density  4  add.gen.  0,1/2,1/2  1/2,0,1/2
SL 10  C(1,1,1/4)       density  8  add.gen.  0,0,1/4  1/2,1/2,0
SL 12  P(1/2,1/2,1/2)   density  8  add.gen.  1/2,0,0  0,1/2,0  0,0,1/2
SL 21  C(1/2,1/2,1/2)   density 16  add.gen.  1/2,0,0  1/4,1/4,0  0,0;1/2
```

Wyckoff letter   non-characteristic crystallographic orbits

```
16 k  1    SL  1  x,y,1/4        P4/mbm (8i,j)
           SL  2  x,1/2+x,z      P4₂/mmc (8o,p)
           SL  3  x,0,z          I4/mmm (16n)
           SL  5  x,1/2+x,1/4    P4/mmm (4l-o)    x,0,1/4      P4/mmm (4j,k)
           SL21  1/4,0,1/4       P4/mmm (1a-d)
 8 j  m    SL  1  x,x,1/4        P4/mbm (4g,h)
           SL  8  1/4,1/4,z      I4/mmm (4e)
           SL12  1/4,1/4,1/4     P4/mmm (1a-d)
 8 i  m    SL  2  x,1/2-x,0      P4₂/mmc (4j-m)   x,1/2+x,0    P4₂/mmc (4j-m)
           SL  3  x,1/2,0        I4/mmm (8i,j)    x,0,0        I4/mmm (8i,j)
 8 h  2    SL  5  0,1/2,z        P4/mmm (2g,h)
           SL10  0,1/2,1/8       P4/mmm (1a-d)
 4 g  mm2  SL  8  1/4,3/4,0      I4/mmm (2a,b)
 4 f  mm2  SL  8  1/4,1/4,0      I4/mmm (2a,b)
 4 e  mm2  SL  3  0,0,z          I4/mmm (4e)
           SL  5  0,0,1/4        P4/mmm (1a-d)
 4 d  4̄    SL  5  0,1/2,1/4      P4/mmm (1a-d)
 4 c  2/m  SL  5  0,1/2,0        P4/mmm (1a-d)
 2 b  mmm  SL  3  0,0,1/2        I4/mmm (2a,b)
 2 a  mmm  SL  3  0,0,0          I4/mmm (2a,b)
```

Space group   No.137   P4$_2$/nmc

Superlattices

```
SL  1  P(1,1,1/2)     density  2  add.gen.  0,0,1/2
SL  2  C(1,1,1)       density  2  add.gen.  1/2,1/2,0
SL  3  I(1,1,1)       density  2  add.gen.  1/2,1/2,1/2
SL  5  C(1,1,1/2)     density  4  add.gen.  0,0,1/2   1/2,1/2,0
SL  6  I(1,1,1/2)     density  4  add.gen.  1/2,1/2,1/4
SL  7  P(1/2,1/2,1)   density  4  add.gen.  1/2,0,0   0,1/2,0
SL 12  P(1/2,1/2,1/2) density  8  add.gen.  1/2,0,0   0,1/2,0   0,0,1/2
SL 15  I(1/2,1/2,1)   density  8  add.gen.  1/2,0,0   1/4,1/4,1/2
SL 19  P(1/2,1/2,1/4) density 16  add.gen.  1/2,0,0   0,1/2,0   0,0,1/4
```

Wyckoff letter   non-characteristic crystallographic orbits

16 h 1	SL 1	x,x,z	P4/nmm (8j)			
	SL 2	x,y,0	P4$_2$/mcm (8n)			
	SL 3	x,y,1/4	I4/mmm (16l)	x,1/2+x,z	I4/mmm (16m)	
	SL 5	x,x,0	P4/mmm (4l-o)	x,x,1/4	P4/mmm (4l-o)	
		x,1/2+x,0	P4/mmm (4l-o)			
	SL 6	x,x,1/8	I4/mmm (8h)			
	SL 7	x,0,1/2	P4$_2$/mmc (4j-m)	x,0,0	P4$_2$/mmc (4j-m)	
	SL12	0,0,z	P4/mmm (2g,h)			
	SL19	0,0,1/8	P4/mmm (1a-d)			
8 g m	SL 2	1/4,y,0	P4$_2$/mcm (4i,j)	1/4,y,1/2	P4$_2$/mcm (4i,j)	
	SL 3	1/4,y,1/4	I4/mmm (8i,j)	1/4,y,3/4	I4/mmm (8i,j)	
	SL15	1/4,0,0	I4/mmm (2a,b)	1/4,0,1/2	I4/mmm (2a,b)	
8 f 2	SL 3	x,-x,1/4	I4/mmm (8h)			
	SL12	0,0,1/4	P4/mmm (1a-d)			
8 e $\bar{1}$	SL12	0,0,0	P4/mmm (1a-d)			
4 d mm2	SL 1	1/4,1/4,z	P4/nmm (2c)			
	SL 5	1/4,1/4,0	P4/mmm (1a-d)	1/4,1/4,1/4	P4/mmm (1a-d)	
	SL 6	1/4,1/4,3/8	I4/mmm (2a,b)	1/4,1/4,1/8	I4/mmm (2a,b)	
4 c mm2	SL 3	3/4,1/4,z	I4/mmm (4e)			
	SL 5	3/4,1/4,0	P4/mmm (1a-d)			
2 b $\bar{4}$m2	SL 3	3/4,1/4,1/4	I4/mmm (2a,b)			
2 a $\bar{4}$m2	SL 3	3/4,1/4,3/4	I4/mmm (2a,b)			

Space group   No.138   P4$_2$/ncm

Superlattices

```
SL  1  P(1,1,1/2)      density  2  add.gen.  0,0,1/2
SL  2  C(1,1,1)        density  2  add.gen.  1/2,1/2,0
SL  3  I(1,1,1)        density  2  add.gen.  1/2,1/2,1/2
SL  5  C(1,1,1/2)      density  4  add.gen.  0,0,1/2  1/2,1/2,0
SL  6  I(1,1,1/2)      density  4  add.gen.  1/2,1/2,1/4
SL  8  F(1,1,1)        density  4  add.gen.  0,1/2,1/2  1/2,0,1/2
SL 10  C(1,1,1/4)      density  8  add.gen.  0,0,1/4  1/2,1/2,0
SL 12  P(1/2,1/2,1/2)  density  8  add.gen.  1/2,0,0  0,1/2,0  0,0,1/2
SL 21  C(1/2,1/2,1/2)  density 16  add.gen.  1/2,0,0  1/4,1/4,0  0,0,1/2
```

Wyckoff letter   non-characteristic crystallographic orbits

16 j 1	SL 1	x,1/4,z	P4/nmm (8i)	
	SL 2	x,1/2+x,z	P4$_2$/mmc (8o,p) x,y,0	P4$_2$/mmc (8q)
	SL 3	x,y,1/4	I4/mcm (16k)	
	SL 5	x,1/2+x,1/4	P4/mmm (4l-o) x,1/4,1/4	P4/mmm (4j,k)
		x,1/4,0	P4/mmm (4j,k)	
	SL 6	x,3/4,1/8	I4/mmm (8i,j) x,1/4,1/8	I4/mmm (8i,j)
	SL 8	x,0,0	I4/mmm (8h)	
	SL21	1/4,0,0	P4/mmm (1a-d) 1/4,0,1/4	P4/mmm (1a-d)
8 i m	SL 2	x,x,1/2	P4$_2$/mmc (4j-m) x,x,0	P4$_2$/mmc (4j-m)
	SL 3	x,x,1/4	I4/mcm (8h)	
	SL 8	0,0,z	I4/mmm (4e)	
	SL12	0,0,1/4	P4/mmm (1a-d)	
8 h 2	SL 2	x,-x,0	P4$_2$/mmc (4j-m)	
8 g 2	SL 2	x,-x,1/2	P4$_2$/mmc (4j-m)	
8 f 2	SL 5	3/4,1/4,h	P4/mmm (2g,h)	
	SL10	3/4,1/4,1/8	P4/mmm (1a-d)	
4 e mm2	SL 1	1/4,1/4,z	P4/nmm (2c)	
	SL 5	1/4,1/4,1/4	P4/mmm (1a-d) 1/4,1/4,0	P4/mmm (1a-d)
	SL 6	1/4,1/4,3/8	I4/mmm (2a,b) 1/4,1/4,1/8	I4/mmm (2a,b)
4 d 2/m	SL 8	0,0,0	I4/mmm (2a,b)	
4 c 2/m	SL 8	0,0,1/2	I4/mmm (2a,b)	
4 b 4̄	SL 5	3/4,1/4,3/4	P4/mmm (1a-d)	
4 a 222	SL 5	3/4,1/4,0	P4/mmm (1a-d)	

Space group   No.139   I4/mmm

Superlattices

```
SL  1  C(1,1,1/2)      density  2  add.gen.  0,0,1/2
SL  2  C(1,1,1/4)      density  4  add.gen.  0,0,1/4
SL  3  P(1/2,1/2,1/2)  density  4  add.gen.  1/2,0,0   0,1/2,0
SL  6  P(1/2,1/2,1/4)  density  8  add.gen.  1/2,0,0   0,0,1/4
SL  8  C(1/2,1/2,1/2)  density  8  add.gen.  1/2,0,0   1/4,1/4,0
```

Wyckoff letter   non-characteristic crystallographic orbits

```
 32 o 1     SL 1   x,y,1/4         P4/mmm (8p,q)   x,1/2+x,z   P4/mmm (8s,t)
            SL 2   x,1/2+x,1/8     P4/mmm (4l-o)
            SL 3   x,1/4,1/4       P4/mmm (4l-o)
 16 n m     SL 1   0,y,1/4         P4/mmm (4j,k)
            SL 8   0,1/4,1/4       P4/mmm (1a-d)
 16 m m     SL 1   x,x,1/4         P4/mmm (4l-o)
            SL 3   1/4,1/4,z       P4/mmm (2g,h)
            SL 6   1/4,1/4,1/8     P4/mmm (1a-d)
 16 l m     SL 1   x,1/2+x,0       P4/mmm (4l-o)
 16 k 2     SL 1   x,1/2+x,1/4     P4/mmm (4l-o)
  8 h mm2   SL 3   1/4,1/4,0       P4/mmm (1a-d)
  8 g mm2   SL 1   0,1/2,z         P4/mmm (2g,h)
            SL 2   0,1/2,1/8       P4/mmm (1a-d)
  8 f 2/m   SL 3   1/4,1/4,1/4     P4/mmm (1a-d)
  4 e 4mm   SL 1   0,0,1/4         P4/mmm (1a-d)
  4 d 4̄m2   SL 1   0,1/2,1/4       P4/mmm (1a-d)
  4 c mmm   SL 1   0,1/2,0         P4/mmm (1a-d)
```

Space group No.140 I4/mcm

Superlattices

```
SL  1   C(1,1,1/2)        density  2   add.gen.   0,0,1/2
SL  2   C(1,1,1/4)        density  4   add.gen.   0,0,1/4
SL  3   P(1/2,1/2,1/2)    density  4   add.gen.   1/2,0,0   0,1/2,0
SL  6   P(1/2,1/2,1/4)    density  8   add.gen.   1/2,0,0   0,0,1/4
SL  8   C(1/2,1/2,1/2)    density  8   add.gen.   1/2,0,0   1/4,1/4,0
SL 13   C(1/2,1/2,1/4)    density 16   add.gen.   1/2,0,0   0,0,1/4   1/4,1/4,0
```

Wyckoff letter   non-characteristic crystallographic orbits

```
32 m 1      SL  1   x,y,1/4        P4/mmm (8p,q)   x,x,z       P4/mmm (8s,t)
                    x,0,z          P4/mmm (8r)
            SL  2   x,x,1/8        P4/mmm (4l-o)   x,0,1/8     P4/mmm (4j,k)
            SL  3   x,1/4,1/4      P4/mmm (4l-o)
            SL  8   1/4,0,z        P4/mmm (2g,h)
            SL 13   1/4,0,1/8      P4/mmm (1a-d)
16 l m      SL  1   x,1/2+x,1/4    P4/mmm (4l-o)
            SL  3   1/4,3/4,z      P4/mmm (2g,h)
            SL  6   1/4,3/4,1/8    P4/mmm (1a-d)
16 k m      SL  1   x,x,0          P4/mmm (4l-o)   x,0,0       P4/mmm (4j,k)
            SL  8   1/4,0,0        P4/mmm (1a-d)
16 j 2      SL  1   x,0,1/4        P4/mmm (4j,k)
            SL  8   1/4,0,1/4      P4/mmm (1a-d)
16 i 2      SL  1   x,x,1/4        P4/mmm (4l-o)
 8 h mm2    SL  3   1/4,3/4,0      P4/mmm (1a-d)
 8 g mm2    SL  1   0,1/2,z        P4/mmm (2g,h)
            SL  2   0,1/2,1/8      P4/mmm (1a-d)
 8 f 4      SL  1   0,0,z          P4/mmm (2g,h)
            SL  2   0,0,1/8        P4/mmm (1a-d)
 8 e 2/m    SL  3   1/4,1/4,1/4    P4/mmm (1a-d)
 4 d mmm    SL  1   0,1/2,0        P4/mmm (1a-d)
 4 c 4/m    SL  1   0,0,0          P4/mmm (1a-d)
 4 b 42m    SL  1   0,1/2,1/4      P4/mmm (1a-d)
 4 a 422    SL  1   0,0,1/4        P4/mmm (1a-d)
```

Space group No.141 I4$_1$/amd

Superlattices

```
SL  3  P(1/2,1/2,1/2)  density  4  add.gen.  1/2,0,0   0,1/2,0
SL  4  F(1,1,1/2)       density  4  add.gen.  0,1/2,1/4
SL  6  P(1/2,1/2,1/4)  density  8  add.gen.  1/2,0,0   0,0,1/4
SL  9  I(1/2,1/2,1/2)  density  8  add.gen.  1/2,0,0   0,1/2,0   1/4,1/4,1/4
SL 11  P(1/2,1/2,1/8)  density 16  add.gen.  1/2,0,0   0,0,1/8
```

Wyckoff letter   non-characteristic crystallographic orbits

```
32 i 1      SL 3   x,0,1/4          P4₂/mmc (4j-m)
            SL 4   x,1/4+x,1/8      I4/mmm (8i,j)
            SL 6   1/4,0,z          P4/mmm (2g,h)
            SL11   1/4,0,1/16       P4/mmm (1a-d)
16 h m      SL 9   0,0,1/4          I4/mmm (2a,b)
16 g 2      SL 6   1/4,1/2,7/8      P4/mmm (1a-d)
16 f 2      SL 6   1/4,0,0          P4/mmm (1a-d)
 8 e mm2    SL 4   0,1/4,1/8        I4/mmm (2a,b)
```

Space group No.142 I4$_1$/acd

Superlattices

```
SL  1  C(1,1,1/2)       density  2  add.gen.  0,0,1/2
SL  3  P(1/2,1/2,1/2)  density  4  add.gen.  1/2,0,0   0,1/2,0
SL  4  F(1,1,1/2)       density  4  add.gen.  0,1/2,1/4
SL  6  P(1/2,1/2,1/4)  density  8  add.gen.  1/2,0,0   0,0,1/4
SL  9  I(1/2,1/2,1/2)  density  8  add.gen.  1/2,0,0   0,1/2,0   1/4,1/4,1/4
SL 11  P(1/2,1/2,1/8)  density 16  add.gen.  1/2,0,0   0,0,1/8
SL 13  C(1/2,1/2,1/4)  density 16  add.gen.  1/2,0,0   0,0,1/4   1/4,1/4,0
```

Wyckoff letter   non-characteristic crystallographic orbits

```
32 g 1      SL 1   x,1/4,z          P4₂/nnm (8m)
            SL 3   x,1/4,0          P4₂/mmc (4j-m)   x,0,0          P4₂/mmc (4j-m)
                   x,1/4,1/4        P4₂/mmc (4j-m)
            SL 4   x,1/4,1/8        I4/mmm (8h)      x,1/4+x,3/8    I4/mmm (8i,j)
            SL 6   1/4,0,z          P4/mmm (2g,h)
            SL 9   1/4,1/4,z        I4/mmm (4e)
            SL11   1/4,0,1/16       P4/mmm (1a-d)
            SL13   0,0,1/8          P4/mmm (1a-d)
16 f 2      SL 6   1/4,1/2,1/8      P4/mmm (1a-d)
16 e 2      SL 6   1/4,0,1/4        P4/mmm (1a-d)
            SL 9   0,0,1/4          I4/mmm (2a,b)
16 d 2      SL 4   0,1/4,z          I4/mmm (4e)
            SL 6   0,1/4,0          P4/mmm (1a-d)
16 c 1̄      SL 9   0,0,0            I4/mmm (2a,b)
 8 b 222    SL 4   0,1/4,1/8        I4/mmm (2a,b)
 8 a 4̄      SL 4   0,1/4,3/8        I4/mmm (2a,b)
```

3.5. Trigonal system

Space group   No.143   P3

Superlattices

```
SL  0  P(1,1,1)          density  1  add.gen.
SL  3  P(1*,1*,1)        density  3  add.gen.  2/3,1/3,0
```

Wyckoff letter   non-characteristic crystallographic orbits

3 d 1	SL 0	x,y,z	P$\bar{6}$ (3j,k)	x,0,z	P$\bar{6}$2m (3f,g)
		x,1/3,z	P$\bar{6}$2m (3f,g)	x,2/3,z	P$\bar{6}$2m (3f,g)
		x,-x,z	P$\bar{6}$m2 (3j,k)	1/2,0,z	P6/mmm (3f,g)
		1/6,1/3,z	P6/mmm (3f,g)	5/6,2/3,z	P6/mmm (3f,g)
	SL 3	1/3,1/3,z	P6/mmm (1a,b)	2/3,2/3,z	P6/mmm (1a,b)
1 c 3	SL 0	2/3,1/3,z	P6/mmm (1a,b)		
1 b 3	SL 0	1/3,2/3,z	P6/mmm (1a,b)		
1 a 3	SL 0	0,0,z	P6/mmm (1a,b)		

Space group   No.144   P3₁

Superlattices

```
SL  0  P(1,1,1)          density  1  add.gen.
SL  2  P(1,1,1/3)        density  3  add.gen.  0,0,1/3
SL  4  RO(1,1,1)         density  3  add.gen.  2/3,1/3,1/3
SL  5  RR(1,1,1)         density  3  add.gen.  1/3,2/3,1/3
```

Wyckoff letter   non-characteristic crystallographic orbits

3 a 1	SL 0	x,0,z	P3₁21 (3a,b)	x,1/3,z	P3₁21 (3a,b)
		x,2/3,z	P3₁21 (3a,b)	x,-x,z	P3₁12 (3a,b)
		1/2,0,z	P6₄22 (3c,d)	1/6,1/3,z	P6₄22 (3c,d)
		5/6,2/3,z	P6₄22 (3c,d)		
	SL 2	2/3,1/3,z	P6/mmm (1a,b)	1/3,2/3,z	P6/mmm (1a,b)
		0,0,z	P6/mmm (1a,b)		
	SL 4	2/3,2/3,z	R$\bar{3}$m (3a,b)		
	SL 5	1/3,1/3,z	R$\bar{3}$m (3a,b)		

Space group   No.145   P3$_2$

Superlattices

```
SL  0   P(1,1,1)          density  1  add.gen.
SL  2   P(1,1,1/3)        density  3  add.gen.  0,0,1/3
SL  4   RO(1,1,1)         density  3  add.gen.  2/3,1/3,1/3
SL  5   RR(1,1,1)         density  3  add.gen.  1/3,2/3,1/3
```

Wyckoff letter   non-characteristic crystallographic orbits

```
   3 a 1    SL 0   x,0,z        P3₂21 (3a,b)    x,1/3,z      P3₂21 (3a,b)
                   x,2/3,z      P3₂21 (3a,b)    x,-x,z       P3₂12 (3a,b)
                   1/2,0,z      P6₂22 (3c,d)    1/6,1/3,z    P6₂22 (3c,d)
                   5/6,2/3,z    P6₂22 (3c,d)
            SL 2   2/3,1/3,z    P6/mmm (1a,b)   1/3,2/3,z    P6/mmm (1a,b)
                   0,0,z        P6/mmm (1a,b)
            SL 4   1/3,1/3,z    R3̄m (3a,b)
            SL 5   2/3,2/3,z    R3̄m (3a,b)
```

Space group   No.146   R3

Superlattices

```
SL  0   RO(1,1,1)         density  1  add.gen.
SL  2   P(1*,1*,1/3)      density  3  add.gen.  0,0,1/3
```

Wyckoff letter   non-characteristic crystallographic orbits

```
   9 b 1    SL 0   x,0,z        R32 (9d,e)      x,-x,z       R3m (9b)
                   1/2,0,z      R3̄m (9d,e)
            SL 2   1/3,1/3,z    P6/mmm (1a,b)   2/3,2/3,z    P6/mmm (1a,b)
   3 a 3    SL 0   0,0,z        R3̄m (3a,b)
```

Space group No.147 P$\bar{3}$

Superlattices

SL	0	P(1,1,1)	density	1	add.gen.	
SL	1	P(1,1,1/2)	density	2	add.gen.	0,0,1/2
SL	3	P(1*,1*,1)	density	3	add.gen.	2/3,1/3,0

Wyckoff letter   non-characteristic crystallographic orbits

6 g 1	SL 0	x,0,z	P$\bar{3}$1m (6k)	x,-x,z	P$\bar{3}$m1 (6i)
		x,y,1/2	P6/m (6j,k)	x,y,0	P6/m (6j,k)
		x,y,1/4	P6$_3$/m (6h)	x,0,0	P6/mmm (6j,k)
		x,0,1/2	P6/mmm (6j,k)	1/2,0,z	P6/mmm (6i)
		x -x,0	P6/mmm (6l,m)	x,-x,1/2	P6/mmm (6l,m)
		x,0,1/4	P6$_3$/mcm (6g)	x,-x,1/4	P6$_3$/mmc (6h)
	SL 1	1/2,0,1,4	P6/mmm (3f,g)		
	SL 3	1/3,0,z	P$\bar{3}$m1 (2d)	1/3,1/3,0	P6/mmm (2c,d)
		1/3,1/3,1/2	P6/ \mm (2c,d)	1/3,1/3,1/4	P6$_3$/mmc (2c,d)
		1/3,1/3,3/4	P6$_3$/mmc (2c,d)		
3 f $\bar{1}$	SL 0	1/2,0,1/2	P6/mmm (3f,g)		
3 e $\bar{1}$	SL 0	1/2,0,0	P6/mmm (3f,g)		
2 d 3	SL 0	1/3,2/3,z	P$\bar{3}$m1 (2d)	1/3,2/3,0	P6/mmm (2c,d)
		1/3,2/3,1/2	P6/mmm (2c,d)	1/3,2/3,1/4	P6$_3$/mmc (2c,d)
		1/3,2/3,3/4	P6$_3$/mmc (2c,d)		
2 c 3	SL 0	0,0,z	P6/mmm (2e)		
	SL 1	0,0,1/4	P6/mmm (1a,b)		
1 b $\bar{3}$	SL 0	0,0,1/2	P6/mmm (1a,b)		
1 a $\bar{3}$	SL 0	0,0,0	P6/mmm (1a,b)		

Space group No.148 R$\bar{3}$

Superlattices

SL	0	RO(1,1,1)	density	1	add.gen.	
SL	1	RR(1,1,1/2)	density	2	add.gen.	0,0,1/2
SL	2	P(1*,1*,1/3)	density	3	add.gen.	0,0,1/3

Wyckoff letter   non-characteristic crystallographic orbits

18 f 1	SL 0	x,-x,z	R$\bar{3}$m (18h)	x,0,1/2	R$\bar{3}$m (18f,g)
		x,0,0	R$\bar{3}$m (18f,g)	x,0,1/4	R$\bar{3}$c (18e)
	SL 1	1/2,0,1/4	R$\bar{3}$m (9d,e)		
	SL 2	1/3,1/3,z	P$\bar{3}$m1 (2d)	1/3,1/3,0	P6/mmm (2c,d)
		1/3,1/3,1/6	P6/mmm (2c,d)	1/3,1/3,1/12	P6$_3$/mmc (2c,d)
		1/3,1/3,3/12	P6$_3$/mmc (2c,d)		
9 e $\bar{1}$	SL 0	1/2,0,0	R$\bar{3}$m (9d,e)		
9 d $\bar{1}$	SL 0	1/2,0,1/2	R$\bar{3}$m (9d,e)		
6 c 3	SL 0	0,0,z	R$\bar{3}$m (6c)		
	SL 1	0,0,1/4	R$\bar{3}$m (3a,b)		
3 b $\bar{3}$	SL 0	0,0,1/2	R$\bar{3}$m (3a,b)		
3 a $\bar{3}$	SL 0	0,0,0	R$\bar{3}$m (3a,b)		

Space group   No.149   P312

Superlattices

```
SL  0  P(1,1,1)        density  1  add.gen.
SL  1  P(1,1,1/2)      density  2  add.gen.  0,0,1/2
SL  3  P(1*,1*,1)      density  3  add.gen.  2/3,1/3,0
```

Wyckoff letter   non-characteristic crystallographic orbits

```
  6 1 1    SL 0   x,y,1/4       P6c2 (6k)        x,y,0         P6m2 (61,m)
                  x,y,1/2       P6m2 (61,m)      x,-x,z        P6m2 (6n)
                  x,0,z         P31m (6k)        x,1/3,z       P31m (6k)
                  x,2/3,z       P31m (6k)        x,0,0         P6/mmm (6j,k)
                  x,1/3,0       P6/mmm (6j,k)    x,2/3,0       P6/mmm (6j,k)
                  x,0,1/2       P6/mmm (6j,k)    x,1/3,1/2     P6/mmm (6j,k)
                  x,2/3,1/2     P6/mmm (6j,k)    1/2,0,z       P6/mmm (6i)
                  5/6,2/3,z     P6/mmm (6i)      1/6,1/3,z     P6/mmm (6i)
                  x,0,1/4       P6₃/mcm (6g)     x,1/3,1/4     P6₃/mcm (6g)
                  x,2/3,1/4     P6₃/mcm (6g)
           SL 1   x,-x,1/4      P6m2 (3j,k)      1/2,1/2,1/4   P6/mmm (3f,g)
                  1/6,5/6,1/4   P6/mmm (3f,g)    5/6,1/6,1/4   P6/mmm (3f,g)
           SL 3   1/3,1/3,z     P3m1 (2d)        1/3,1/3,0     P6/mmm (2c,d)
                  1/3,1/3,1/2   P6/mmm (2c,d)    1/3,1/3,1/4   P6₃/mmc (2c,d)
                  1/3,1/3,3/4   P6₃/mmc (2c,d)
  3 k 2    SL 0   x,-x,1/2      P6m2 (3j,k)      1/2,1/2,1/2   P6/mmm (3f,g)
                  5/6,1/6,1/2   P6/mmm (3f,g)    1/6,5/6,1/2   P6/mmm (3f,g)
  3 j 2    SL 0   x,-x,0        P6m2 (3j,k)      1/2,1/2,0     P6/mmm (3f,g)
                  5/6,1/6,0     P6/mmm (3f,g)    1/6,5/6,0     P6/mmm (3f,g)
  2 i 3    SL 0   2/3,1/3,z     P6/mmm (2e)
           SL 1   2/3,1/3,1/4   P6/mmm (1a,b)
  2 h 3    SL 0   1/3,2/3,z     P6/mmm (2e)
           SL 1   1/3,2/3,1/4   P6/mmm (1a,b)
  2 g 3    SL 0   0,0,z         P6/mmm (2e)
           SL 1   0,0,1/4       P6/mmm (1a,b)
  1 f 32   SL 0   2/3,1/3,1/2   P6/mmm (1a,b)
  1 e 32   SL 0   2/3,1/3,0     P6/mmm (1a,b)
  1 d 32   SL 0   1/3,2/3,1/2   P6/mmm (1a,b)
  1 c 32   SL 0   1/3,2/3,0     P6/mmm (1a,b)
  1 b 32   SL 0   0,0,1/2       P6/mmm (1a,b)
  1 a 32   SL 0   0,0,0         P6/mmm (1a,b)
```

Space group No.150 P321

Superlattices

```
SL  0  P(1,1,1)      density  1  add.gen.
SL  1  P(1,1,1/2)    density  2  add.gen.  0,0,1/2
SL  3  P(1*,1*,1)    density  3  add.gen.  2/3,1/3,0
SL  9  P(1*,1*,1/2)  density  6  add.gen.  0,0,1/2  2/3,1/3,0
```

Wyckoff letter   non-characteristic crystallographic orbits

```
6 g 1    SL 0  x,y,1/2       P6̄2m (6j,k)     x,y,0         P6̄2m (6j,k)
               x,y,1/4       P6̄2c (6h)       x,0,z         P6̄2m (6i)
               x,-x,z        P3̄m1 (6i)       x,-x,0        P6/mmm (6l,m)
               x,-x,1/2      P6/mmm (6l,m)   x,-x,1/4      P6₃/mmc (6h)
               1/2,0,z       P6/mmm (6i)
         SL 1  x,0,1/4       P6̄2m (3f,g)     1/2,0,1/4     P6/mmm (3f,g)
         SL 3  2/3,2/3,z     P6/mmm (2e)     1/3,1/3,z     P6/mmm (2e)
         SL 9  2/3,2/3,1/4   P6/mmm (1a,b)   1/3,1/3,1/4   P6/mmm (1a,b)
3 f 2    SL 0  x,0,1/2       P6̄2m (3f,g)     1/2,0,1/2     P6/mmm (3f,g)
         SL 3  1/3,0,1/2     P6/mmm (1a,b)   2/3,0,1/2     P6/mmm (1a,b)
3 e 2    SL 0  x,0,0         P6̄2m (3f,g)     1/2,0,0       P6/mmm (3f,g)
         SL 3  1/3,0,0       P6/mmm (1a,b)   2/3,0,0       P6/mmm (1a,b)
2 d 3    SL 0  1/3,2/3,z     P3̄m1 (2d)       1/3,2/3,1/2   P6/mmm (2c,d)
               1/3,2/3,0     P6/mmm (2c,d)   1/3,2/3,1/4   P6₃/mmc (2c,d)
               1/3,2/3,3/4   P6₃/mmc (2c,d)
2 c 3    SL 0  0,0,z         P6/mmm (2e)
         SL 1  0,0,1/4       P6/mmm (1a,b)
1 b 32   SL 0  0,0,1/2       P6/mmm (1a,b)
1 a 32   SL 0  0,0,0         P6/mmm (1a,b)
```

Space group No.151 $P3_1 12$

Superlattices

SL 0	P(1,1,1)	density	1	add.gen.	
SL 1	P(1,1,1/2)	density	2	add.gen.	0,0,1/2
SL 2	P(1,1,1/3)	density	3	add.gen.	0,0,1/3
SL 8	P(1,1,1/6)	density	6	add.gen.	0,0,1/6

Wyckoff letter non-characteristic crystallographic orbits

6 c 1	SL 0	x,0,5/12	$P6_1 22$ (6a)	x,1/3,5/12	$P6_1 22$ (6a)
		x,2/3,5/12	$P6_1 22$ (6a)	x,0,1/6	$P6_4 22$ (6g,h)
		x,1/3,1/6	$P6_4 22$ (6g,h)	x,2/3,1/6	$P6_4 22$ (6g,h)
		x,0,2/3	$P6_4 22$ (6g,h)	x,1/3,2/3	$P6_4 22$ (6g,h)
		x,2/3,2/3	$P6_4 22$ (6g,h)	1/2,0,z	$P6_4 22$ (6f)
		1/6,1/3,z	$P6_4 22$ (6f)	5/6,2/3,z	$P6_4 22$ (6f)
	SL 1	x,-x,1/12	$P3_2 12$ (3a,b)	1/2,1/2,1/12	$P6_2 22$ (3c,d)
		5/6,1/6,1/12	$P6_2 22$ (3c,d)	1/6,5/6,1/12	$P6_2 22$ (3c,d)
	SL 2	2/3,1/3,z	P6/mmm (2e)	1/3,2/3,z	P6/mmm (2e)
		0,0,z	P6/mmm (2e)		
	SL 8	1/3,2/3,1/12	P6/mmm (1a,b)	0,0,1/12	P6/mmm (1a,b)
		2/3,1/3,1/12	P6/mmm (1a,b)		
3 b 2	SL 0	1/2,1/2,5/6	$P6_4 22$ (3c,d)	1/6,5/6,5/6	$P6_4 22$ (3c,d)
		5/6,1/6,5/6	$P6_4 22$ (3c,d)		
	SL 2	2/3,1/3,5/6	P6/mmm (1a,b)	1/3,2/3,5/6	P6/mmm (1a,b)
		0,0,5/6	P6/mmm (1a,b)		
3 a 2	SL 0	1/2,1/2,1/3	$P6_4 22$ (3c,d)	1/6,5/6,1/3	$P6_4 22$ (3c,d)
		5/6,1/6,1/3	$P6_4 22$ (3c,d)		
	SL 2	2/3,1/3,1/3	P6/mmm (1a,b)	1/3,2/3,1/3	P6/mmm (1a,b)
		0,0,1/3	P6/mmm (1a,b)		

Space group No.152 $P3_121$

Superlattices

SL	0	P(1,1,1)	density	1	add.gen.	
SL	1	P(1,1,1/2)	density	2	add.gen.	0,0,1/2
SL	2	P(1,1,1/3)	density	3	add.gen.	0,0,1/3
SL	4	RO(1,1,1)	density	3	add.gen.	2/3,1/3,1/3
SL	5	RR(1,1,1)	density	3	add.gen.	1/3,2/3,1/3
SL	8	P(1,1,1/6)	density	6	add.gen.	0,0,1/6
SL	10	RO(1,1,1/2)	density	6	add.gen.	2/3,1/3,1/6
SL	11	RR(1,1,1/2)	density	6	add.gen.	1/3,2/3,1/6

Wyckoff letter		non-characteristic crystallographic orbits			
6 c 1	SL 0	x,-x,3/12	$P6_122$ (6b)	x,-x,1/2	$P6_422$ (6i,j)
		x,-x,0	$P6_422$ (6i,j)	1/2,0,z	$P6_422$ (6f)
	SL 1	x,0,1/12	$P3_221$ (3a,b)	1/2,0,1/12	$P6_222$ (3c,d)
	SL 2	2/3,1/3,z	$P\bar{3}m1$ (2d)	2/3,1/3,1/6	P6/mmm (2c,d)
		2/3,1/3,0	P6/mmm (2c,d)	2/3,1/3,1/12	$P6_3/mmc$ (2c,d)
		2/3,1/3,3/12	$P6_3/mmc$ (2c,d)	0,0,z	P6/mmm (2e)
	SL 4	2/3,2/3,z	$R\bar{3}m$ (6c)		
	SL 5	1/3,1/3,z	$R\bar{3}m$ (6c)		
	SL 8	0,0,1/12	P6/mmm (1a,b)		
	SL10	1/3,1/3,1/4	$R\bar{3}m$ (3a,b)		
	SL11	2/3,2/3,1/4	$R\bar{3}m$ (3a,b)		
3 b 2	SL 0	1/2,0,5/6	$P6_422$ (3c,d)		
	SL 2	0,0,5/6	P6/mmm (1a,b)		
	SL 4	1/3,0,5/6	$R\bar{3}m$ (3a,b)		
	SL 5	2/3,0,5/6	$R\bar{3}m$ (3a,b)		
3 a 2	SL 0	1/2,0,1/3	$P6_422$ (3c,d)		
	SL 2	0,0,1/3	P6/mmm (1a,b)		
	SL 4	1/3,0,1/3	$R\bar{3}m$ (3a,b)		
	SL 5	2/3,0,1/3	$R\bar{3}m$ (3a,b)		

Space group No.153 P3$_2$12

Superlattices

```
SL  0  P(1,1,1)        density  1  add.gen.
SL  1  P(1,1,1/2)      density  2  add.gen.  0,0,1/2
SL  2  P(1,1,1/3)      density  3  add.gen.  0,0,1/3
SL  8  P(1,1,1/6)      density  6  add.gen.  0,0,1/6
```

Wyckoff letter   non-characteristic crystallographic orbits

6 c 1	SL 0	x,0,1/12	P6$_5$22 (6a)		x,1/3,1/12	P6$_5$22 (6a)
		x,2/3,1/12	P6$_5$22 (6a)		x,0,1/3	P6$_2$22 (6g,h)
		x,1/3,1/3	P6$_2$22 (6g,h)		x,2/3,1/3	P6$_2$22 (6g,h)
		x,0,5/6	P6$_2$22 (6g,h)		x,1/3,5/6	P6$_2$22 (6g,h)
		x,2/3,5/6	P6$_2$22 (6g,h)		1/2,0,z	P6$_2$22 (6f)
		1/6,1/3,z	P6$_2$22 (6f)		5/6,2/3,z	P6$_2$22 (6f)
	SL 1	x,-x,5/12	P3$_1$12 (3a,b)		1/2,1/2,5/12	P6$_4$22 (3c,d)
		5/6,1/6,5/12	P6$_4$22 (3c,d)		1/6,5/6,5/12	P6$_4$22 (3c,d)
	SL 2	2/3,1/3,z	P6/mmm (2e)		1/3,2/3,z	P6/mmm (2e)
		0,0,z	P6/mmm (2e)			
	SL 8	1/3,2/3,1/12	P6/mmm (1a,b)		0,0,1/12	P6/mmm (1a,b)
		2/3,1/3,1/12	P6/mmm (1a,b)			
3 b 2	SL 0	1/2,1/2,1/6	P6$_2$22 (3c,d)		1/6,5/6,1/6	P6$_2$22 (3c,d)
		5/6,1/6,1/6	P6$_2$22 (3c,d)			
	SL 2	2/3,1/3,1/6	P6/mmm (1a,b)		1/3,2/3,1/6	P6/mmm (1a,b)
		0,0,1/6	P6/mmm (1a,b)			
3 a 2	SL 0	1/2,1/2,2/3	P6$_2$22 (3c,d)		1/6,5/6,2/3	P6$_2$22 (3c,d)
		5/6,1/6,2/3	P6$_2$22 (3c,d)			
	SL 2	2/3,1/3,2/3	P6/mmm (1a,b)		1/3,2/3,2/3	P6/mmm (1a,b)
		0,0,2/3	P6/mmm (1a,b)			

Space group No.154 P3$_2$21

Superlattices

SL	0	P(1,1,1)	density	1	add.gen.	
SL	1	P(1,1,1/2)	density	2	add.gen.	0,0,1/2
SL	2	P(1,1,1/3)	density	3	add.gen.	0,0,1/3
SL	4	RO(1,1,1)	density	3	add.gen.	2/3,1/3,1/3
SL	5	RR(1,1,1)	density	3	add.gen.	1/3,2/3,1/3
SL	8	P(1,1,1/6)	density	6	add.gen.	0,0,1/6
SL	10	RO(1,1,1/2)	density	6	add.gen.	2/3,1/3,1/6
SL	11	RR(1,1,1/2)	density	6	add.gen.	1/3,2/3,1/6

Wyckoff letter   non-characteristic crystallographic orbits

Wyckoff	SL	coord	orbit	coord	orbit
6 c 1	SL 0	x,-x,1/4	P6$_5$22 (6b)	x,-x,0	P6$_2$22 (6i,j)
		x,-x,1/2	P6$_2$22 (6i,j)	1/2,0,z	P6$_2$22 (6f)
	SL 1	x,0,5/12	P3$_1$21 (3a,b)	1/2,0,5/12	P6$_4$22 (3c,d)
	SL 2	2/3,1/3,z	P$\bar{3}$m1 (2d)	2/3,1/3,1/6	P6/mmm (2c,d)
		2/3,1/3,0	P6/mmm (2c,d)	2/3,1/3,3/4	P6$_3$/mmc (2c,d)
		2/3,1/3,1/4	P6$_3$/mmc (2c,d)	0,0,z	P6/mmm (2e)
	SL 4	1/3,1/3,z	R$\bar{3}$m (6c)		
	SL 5	2/3,2/3,z	R$\bar{3}$m (6c)		
	SL 8	0,0,1/12	P6/mmm (1a,b)		
	SL10	2/3,2/3,1/4	R$\bar{3}$m (3a,b)		
	SL11	1/3,1/3,1/4	R$\bar{3}$m (3a,b)		
3 b 2	SL 0	1/2,0,1/6	P6$_2$22 (3c,d)		
	SL 2	0,0,1/6	P6/mmm (1a,b)		
	SL 4	2/3,0,1/6	R$\bar{3}$m (3a,b)		
	SL 5	1/3,0,1/6	R$\bar{3}$m (3a,b)		
3 a 2	SL 0	1/2,0,2/3	P6$_2$22 (3c,d)		
	SL 2	0,0,2/3	P6/mmm (1a,b)		
	SL 4	2/3,0,2/3	R$\bar{3}$m (3a,b)		
	SL 5	1/3,0,2/3	R$\bar{3}$m (3a,b)		

Space group No.155 R32

Superlattices

```
SL  0  RO(1,1,1)        density  1  add.gen.
SL  1  RR(1,1,1/2)      density  2  add.gen.  0,0,1/2
SL  2  P(1*,1*,1/3)     density  3  add.gen.  0,0,1/3
SL  5  P(1*,1*,1/6)     density  6  add.gen.  0,0,1/6
```

Wyckoff letter   non-characteristic crystallographic orbits

```
 18 f 1    SL 0  x,-x,z          R3̄m (18h)
           SL 1  x,0,1/4         R32 (9d,e)      1/2,0,1/4     R3̄m (9d,e)
           SL 2  1/3,1/3,z       P6/mmm (2e)     2/3,2/3,z     P6/mmm (2e)
           SL 5  1/3,1/3,1/12    P6/mmm (1a,b)   2/3,2/3,1/12  P6/mmm (1a,b)
  9 e 2    SL 0  1/2,0,1/2       R3̄m (9d,e)
           SL 2  2/3,0,1/2       P6/mmm (1a,b)   1/3,0,1/2     P6/mmm (1a,b)
  9 d 2    SL 0  1/2,0,0         R3̄m (9d,e)
           SL 2  2/3,0,0         P6/mmm (1a,b)   1/3,0,0       P6/mmm (1a,b)
  6 c 3    SL 0  0,0,z           R3̄m (6c)
           SL 1  0,0,1/4         R3̄m (3a,b)
  3 b 32   SL 0  0,0,1/2         R3̄m (3a,b)
  3 a 32   SL 0  0,0,0           R3̄m (3a,b)
```

Space group No.156 P3m1

Superlattices

```
SL  0  P(1,1,1)         density  1  add.gen.
SL  3  P(1*,1*,1)       density  3  add.gen.  2/3,1/3,0
```

Wyckoff letter   non-characteristic crystallographic orbits

```
  6 e 1    SL 0  x,y,z           P6̄m2 (61,m)    x,0,z         P6/mmm (6j,k)
                 x,1/3,z         P6/mmm (6j,k)  x,2/3,z       P6/mmm (6j,k)
           SL 3  1/3,1/3,z       P6/mmm (2c,d)
  3 d m    SL 0  x,-x,z          P6̄m2 (3j,k)    1/2,1/2,z     P6/mmm (3f,g)
                 1/6,5/6,z       P6/mmm (3f,g)  5/6,1/6,z     P6/mmm (3f,g)
  1 c 3m   SL 0  2/3,1/3,z       P6/mmm (1a,b)
  1 b 3m   SL 0  1/3,2/3,z       P6/mmm (1a,b)
  1 a 3m   SL 0  0,0,z           P6/mmm (1a,b)
```

Space group   No.157   P31m

Superlattices

```
SL  0   P(1,1,1)          density  1  add.gen.
SL  3   P(1*,1*,1)        density  3  add.gen.   2/3,1/3,0
```

Wyckoff letter   non-characteristic crystallographic orbits

```
  6 d 1      SL 0   x,y,z       P6̄2m (6j,k)      x,-x,z      P6/mmm (6l,m)
  3 c m      SL 0   x,0,z       P6̄2m (3f,g)      1/2,0,z     P6/mmm (3f,g)
             SL 3   2/3,0,z     P6/mmm (1a,b)    1/3,0,z     P6/mmm (1a,b)
  2 b 3      SL 0   1/3,2/3,z   P6/mmm (2c,d)
  1 a 3m     SL 0   0,0,z       P6/mmm (1a,b)
```

Space group   No.158   P3c1

Superlattices

```
SL  0   P(1,1,1)          density  1  add.gen.
SL  1   P(1,1,1/2)        density  2  add.gen.   0,0,1/2
SL  3   P(1*,1*,1)        density  3  add.gen.   2/3,1/3,0
```

Wyckoff letter   non-characteristic crystallographic orbits

```
  6 d 1      SL 0   x,y,z       P6̄c2 (6k)        x,0,z       P6₃/mcm (6g)
                    x,1/3,z     P6₃/mcm (6g)     x,2/3,z     P6₃/mcm (6g)
             SL 1   x,-x,z      P6̄m2 (3j,k)      1/2,1/2,z   P6/mmm (3f,g)
                    1/6,5/6,z   P6/mmm (3f,g)    5/6,1/6,z   P6/mmm (3f,g)
             SL 3   1/3,1/3,z   P6₃/mmc (2c,d)
  2 c 3      SL 1   2/3,1/3,z   P6/mmm (1a,b)
  2 b 3      SL 1   1/3,2/3,z   P6/mmm (1a,b)
  2 a 3      SL 1   0,0,z       P6/mmm (1a,b)
```

Space group No.159 P31c

Superlattices

```
SL  0  P(1,1,1)        density  1  add.gen.
SL  1  P(1,1,1/2)      density  2  add.gen.  0,0,1/2
SL  9  P(1*,1*,1/2)    density  6  add.gen¦  0,0,1/2  2/3,1/3,0
```

Wyckoff letter   non-characteristic crystallographic orbits

```
   6 c 1      SL 0  x,y,z      P6̄2c (6h)        x,-x,z      P6₃/mmc (6h)
              SL 1  x,0,z      P6̄2m (3f,g)      1/2,0,z     P6/mmm (3f,g)
              SL 9  1/3,1/3,z  P6/mmm (1a,b)    2/3,2/3,z   P6/mmm (1a,b)
   2 b 3      SL 0  1/3,2/3,z  P6₃/mmc (2c,d)
   2 a 3      SL 1  0,0,z      P6/mmm (1a,b)
```

Space group No.160 R3m

Superlattices

```
SL  0  RO(1,1,1)       density  1  add.gen.
SL  2  P(1*,1*,1/3)    density  3  add.gen.  0,0,1/3
```

Wyckoff letter   non-characteristic crystallographic orbits

```
  18 c 1      SL 0  x,0,z      R3̄m (18f,g)
              SL 2  1/3,1/3,z  P6/mmm (2c,d)
   9 b m      SL 0  1/2,1/2,z  R3̄m (9d,e)
   3 a 3m     SL 0  0,0,z      R3̄m (3a,b)
```

Space group   No.161   R3c

Superlattices

```
SL  0   RO(1,1,1)          density  1  add.gen.
SL  1   RR(1,1,1/2)        density  2  add.gen.   0,0,1/2
SL  2   P(1*,1*,1/3)       density  3  add.gen.   0,0,1/3
```

Wyckoff letter   non-characteristic crystallographic orbits

```
  18 b 1      SL 0   x,0,z        R3̄c (18e)
              SL 1   x,-x,z       R3m (9b)           1/2,0,z      R3̄m (9d,e)
              SL 2   1/3,1/3,z    P6₃/mmc (2c,d)
   6 a 3      SL 1   0,0,z        R3̄m (3a,b)
```


Space group   No.162   P3̄1m

Superlattices

```
SL  0   P(1,1,1)           density  1  add.gen.
SL  1   P(1,1,1/2)         density  2  add.gen.   0,0,1/2
SL  3   P(1*,1*,1)         density  3  add.gen.   2/3,1/3,0
```

Wyckoff letter   non-characteristic crystallographic orbits

```
  12 l 1     SL 0   x,y,1/2       P6/mmm (12p,q)   x,y,1/4     P6₃/mcm (12j)
                    x,y,0         P6/mmm (12p,q)   x,2x,z      P6/mmm (12o)
             SL 1   x,-x,1/4      P6/mmm (6l,m)
   6 k m     SL 0   x,0,1/2       P6/mmm (6j,k)    x,0,1/4     P6₃/mcm (6g)
                    x,0,0         P6/mmm (6j,k)    1/2,0,z     P6/mmm (6l)
             SL 1   1/2,0,1/4     P6/mmm (3f,g)
             SL 3   2/3,0,z       P3̄m1 (2d)        2/3,0,0     P6/mmm (2c,d)
                    2/3,0,1/2     P6/mmm (2c,d)    2/3,0,1/4   P6₃/mmc (2c,d
                    2/3,0,3/4     P6₃/mmc (2c,d)
   6 j 2     SL 0   x,-x,1/2      P6/mmm (6l,m)
   6 i 2     SL 0   x,-x,0        P6/mmm (6l,m)
   4 h 3     SL 0   1/3,2/3,z     P6/mmm (4h)
             SL 1   1/3,2/3,1/4   P6/mmm (2c,d)
   3 g 2/m   SL 0   1/2,0,1/2     P6/mmm (3f,g)
   3 f 2/m   SL 0   1/2,0,0       P6/mmm (3f,g)
   2 e 3m    SL 0   0,0,z         P6/mmm (2e)
             SL 1   0,0,1/4       P6/mmm (1a,b)
   2 d 32    SL 0   1/3,2/3,1/2   P6/mmm (2c,d)
   2 c 32    SL 0   1/3,2/3,0     P6/mmm (2c,d)
   1 b 3̄m    SL 0   0,0,1/2       P6/mmm (1a,b)
   1 a 3̄m    SL 0   0,0,0         P6/mmm (1a,b)
```

Space group No.163 P$\bar{3}$1c

Superlattices

SL	0	P(1,1,1)	density	1	add.gen.		
SL	1	P(1,1,1/2)	density	2	add.gen.	0,0,1/2	
SL	6	P(1,1,1/4)	density	4	add.gen.	0,0,1/4	
SL	9	P(1*,1*,1/2)	density	6	add.gen.	0,0,1/2	2/3,1/3,0

Wyckoff letter   non-characteristic crystallographic orbits

12 i 1	SL 0	x,y,0	P6/mcc (121)	x,y,1/4	P6$_3$/mmc (12j)
		x,2x,z	P6$_3$/mmc (12k)		
	SL 1	x,0,z	P$\bar{3}$1m (6k)	x,0,0	P6/mmm (6j,k)
		x,0,1/4	P6/mmm (6j,k)	1/2,0,z	P6/mmm (6i)
		x,0,1/8	P6$_3$/mcm (6g)	x,2x,0	P6/mmm (6l,m)
	SL 6	1/2,1/2,1/8	P6/mmm (3f,g)		
	SL 9	1/3,1/3,z	P$\bar{3}$m1 (2d)	1/3,1/3,0	P6/mmm (2c,d)
		1/3,1/3,1/4	P6/mmm (2c,d)	1/3,1/3,1/8	P6$_3$/mmc (2c,d)
		1/3,1/3,3/8	P6$_3$/mmc (2c,d)		
6 h 2	SL 0	x,-x,1/4	P6$_3$/mmc (6h)		
	SL 1	1/2,1/2,1/4	P6/mmm (3f,g)		
6 g $\bar{1}$	SL 1	1/2,0,0	P6/mmm (3f,g)		
4 f 3	SL 0	1/3,2/3,z	P6$_3$/mmc (4f)		
	SL 1	1/3,2/3,0	P6/mmm (2c,d)		
4 e 3	SL 1	0,0,z	P6/mmm (2e)		
	SL 6	0,0,1/8	P6/mmm (1a,b)		
2 d 32	SL 0	2/3,1/3,1/4	P6$_3$/mmc (2c,d)		
2 c 32	SL 0	1/3,2/3,1/4	P6$_3$/mmc (2c,d)		
2 b $\bar{3}$	SL 1	0,0,0	P6/mmm (1a,b)		
2 a 32	SL 1	0,0,1/4	P6/mmm (1a,b)		

Space group   No.164   P3̄m1

Superlattices

```
SL  0   P(1,1,1)        density  1  add.gen.
SL  1   P(1,1,1/2)      density  2  add.gen.  0,0,1/2
SL  3   P(1*,1*,1)      density  3  add.gen.  2/3,1/3,0
SL  9   P(1*,1*,1/2)    density  6  add.gen.  0,0,1/2  2/3,1/3,0
```

Wyckoff letter   non-characteristic crystallographic orbits

12 j 1	SL 0	x,y,1/2	P6/mmm (12p,q)	x,y,1/4	P6₃/mmc (12j)	
		x,y,0	P6/mmm (12p,q)	x,0,z	P6/mmm (12n)	
	SL 1	x,0,1/4	P6/mmm (6j,k)			
	SL 3	1/3,1/3,z	P6/mmm (4h)			
	SL 9	1/3,1/3,1/4	P6/mmm (2c,d)			
6 i m	SL 0	x,-x,1/2	P6/mmm (6l,m)	x,-x,1/4	P6₃/mmc (6h)	
		x,-x,0	P6/mmm (6l,m)	1/2,1/2,z	P6/mmm (6i)	
	SL 1	1/2,1/2,1/4	P6/mmm (3f,g)			
6 h 2	SL 0	x,0,1/2	P6/mmm (6j,k)			
	SL 3	1/3,0,1/2	P6/mmm (2c,d)			
6 g 2	SL 0	x,0,0	P6/mmm (6j,k)			
	SL 3	1/3,0,0	P6/mmm (2c,d)			
3 f 2/m	SL 0	1/2,0,1/2	P6/mmm (3f,g)			
3 e 2/m	SL 0	1/2,0,0	P6/mmm (3f,g)			
2 d 3m	SL 0	1/3,2/3,1/2	P6/mmm (2c,d)	1/3,2/3,1/4	P6₃/mmc (2c,d)	
		1/3,2/3,3/4	P6₃/mmc (2c,d)	1/3,2/3,0	P6/mmm (2c,d)	
2 c 3m	SL 0	0,0,z	P6/mmm (2e)			
	SL 1	0,0,1/4	P6/mmm (1a,b)			
1 b 3̄m	SL 0	0,0,1/2	P6/mmm (1a,b)			
1 a 3̄m	SL 0	0,0,0	P6/mmm (1a,b)			

Space group   No.165 P3̄c1

Superlattices

```
SL  0  P(1,1,1)        density  1  add.gen.
SL  1  P(1,1,1/2)      density  2  add.gen.  0,0,1/2
SL  3  P(1*,1*,1)      density  3  add.gen.  2/3,1/3,0
SL  6  P(1,1,1/4)      density  4  add.gen.  0,0,1/4
SL  9  P(1*,1*,1/2)    density  6  add.gen.  0,0,1/2   2/3,1/3,0
```

Wyckoff letter   non-characteristic crystallographic orbits

```
 12 g 1    SL 0   x,y,0        P6/mcc (12l)    x,0,z        P6₃/mcm (12k)
                  x,y,1/4      P6₃/mcm (12j)
           SL 1   x,-x,z       P3̄m1 (6i)      x,-x,0       P6/mmm (6l,m)
                  x,-x,1/4     P6/mmm (6l,m)   x,-x,1/8     P6₃/mmc (6h)
                  1/2,0,z      P6/mmm (6i)     x,0,0        P6/mmm (6j,k)
           SL 3   1/3,1/3,z    P6₃/mmc (4f)
           SL 6   1/2,0,1/8    P6/mmm (3f,g)
           SL 9   1/3,1/3,0    P6/mmm (2c,d)
  6 f 2    SL 0   x,0,1/4      P6₃/mcm (6g)
           SL 1   1/2,0,1/4    P6/mmm (3f,g)
           SL 3   2/3,0,1/4    P6₃/mmc (2c,d)  1/3,0,1/4    P6₃/mmc (2c,d)
  6 e 1̄    SL 1   1/2,0,0      P6/mmm (3f,g)
  4 d 3    SL 1   1/3,2/3,z    P3̄m1 (2d)      1/3,2/3,0    P6/mmm (2c,d)
                  1/3,2/3,1/4  P6/mmm (2c,d)   1/3,2/3,1/8  P6₃/mmc (2c,d)
                  1/3,2/3,3/8  P6₃/mmc (2c,d)
  4 c 3    SL 1   0,0,z        P6/mmm (2e)
           SL 6   0,0,1/8      P6/mmm (1a,b)
  2 b 3̄    SL 1   0,0,0        P6/mmm (1a,b)
  2 a 32   SL 1   0,0,1/4      P6/mmm (1a,b)
```

Space group   No.166   R$\bar{3}$m

Superlattices

```
SL  1  RR(1,1,1/2)     density  2  add.gen.   0,0,1/2
SL  2  P(1*,1*,1/3)    density  3  add.gen.   0,0,1/3
SL  5  P(1*,1*,1/6)    density  6  add.gen.   0,0,1/6
```

Wyckoff letter   non-characteristic crystallographic orbits

```
  36 i 1      SL 1  x,0,1/4        R̄3m (18f,g)
              SL 2  1/3,1/3,z      P6/mmm (4h)
              SL 5  1/3,1/3,1/12   P6/mmm (2c,d)
  18 h m      SL 1  1/2,1/2,1/4    R̄3m (9d,e)
  18 g 2      SL 2  1/3,0,1/2      P6/mmm (2c,d)
  18 f 2      SL 2  1/3,0,0        P6/mmm (2c,d)
   6 c 3m     SL 1  0,0,1/4        R̄3m (3a,b)
```

Space group   No.167   R$\bar{3}$c

Superlattices

```
SL  1  RR(1,1,1/2)     density  2  add.gen.   0,0,1/2
SL  2  P(1*,1*,1/3)    density  3  add.gen.   0,0,1/3
SL  3  RO(1,1,1/4)     density  4  add.gen.   0,0,1/4
SL  5  P(1*,1*,1/6)    density  6  add.gen.   0,0,1/6
```

Wyckoff letter   non-characteristic crystallographic orbits

```
  36 f 1      SL 1  x,-x,z      R̄3m (18h)          x,0,0     R̄3m (18f,g)
              SL 2  1/3,1/3,z   P6₃/mmc (4f)
              SL 3  1/2,0,1/8   R̄3m (9d,e)
              SL 5  1/3,1/3,0   P6/mmm (2c,d)
  18 e 2      SL 1  1/2,0,1/4   R̄3m (9d,e)
              SL 2  2/3,0,1/4   P6₃/mmc (2c,d)   1/3,0,1/4   P6₃/mmc (2c,d)
  18 d 1̄     SL 1  1/2,0,0     R̄3m (9d,e)
  12 c 3      SL 1  0,0,z       R̄3m (6c)
              SL 3  0,0,1/8     R̄3m (3a,b)
   6 b 3̄     SL 1  0,0,0       R̄3m (3a,b)
   6 a 32     SL 1  0,0,1/4     R̄3m (3a,b)
```

3.6. Hexagonal system

Space group No.168 P6

Superlattices

```
SL  0  P(1,1,1)        density  1  add.gen.
SL  3  P(1*,1*,1)      density  3  add.gen.  2/3,1/3,0
```

Wyckoff letter   non-characteristic crystallographic orbits

```
    6 d 1     SL 0   x,y,z       P6/m (6j,k)      x,0,z       P6/mmm (6j,k)
                     x,-x,z      P6/mmm (6l,m)
              SL 3   1/3,1/3,z   P6/mmm (2c,d)
    3 c 2     SL 0   1/2,0,z     P6/mmm (3f,g)
    2 b 3     SL 0   1/3,2/3,z   P6/mmm (2c,d)
    1 a 6     SL 0   0,0,z       P6/mmm (1a,b)
```

Space group No.169 P6$_1$

Superlattices

```
SL  0  P(1,1,1)        density  1  add.gen.
SL  1  P(1,1,1/2)      density  2  add.gen.  0,0,1/2
SL  2  P(1,1,1/3)      density  3  add.gen.  0,0,1/3
SL  6  P(1,1,1/6)      density  6  add.gen.  0,0,1/6
```

Wyckoff letter   non-characteristic crystallographic orbits

```
    6 a 1     SL 0   x,-x,z      P6₁22 (6b)       x,0,z       P6₁22 (6a)
              SL 1   1/2,0,z     P6₂22 (3c,d)
              SL 2   2/3,1/3,z   P6₃/mmc (2c,d)
              SL 6   0,0,z       P6/mmm (1a,b)
```

Space group   No.170   P6$_5$

Superlattices

```
SL  0   P(1,1,1)        density  1  add.gen.
SL  1   P(1,1,1/2)      density  2  add.gen.   0,0,1/2
SL  2   P(1,1,1/3)      density  3  add.gen.   0,0,1/3
SL  6   P(1,1,1/6)      density  6  add.gen.   0,0,1/6
```

Wyckoff letter   non-characteristic crystallographic orbits

```
   6 a 1    SL 0  x,-x,z      P6₅22 (6b)       x,0,z        P6₅22 (6a)
            SL 1  1/2,0,z     P6₄22 (3c,d)
            SL 2  2/3,1/3,z   P6₃/mmc (2c,d)
            SL 6  0,0,z       P6/mmm (1a,b)
```

Space group   No.171   P6$_2$

Superlattices

```
SL  0   P(1,1,1)        density  1  add.gen.
SL  2   P(1,1,1/3)      density  3  add.gen.   0,0,1/3
```

Wyckoff letter   non-characteristic crystallographic orbits

```
   6 c 1    SL 0  x,-x,z      P6₂22 (6i,j)     x,0,z        P6₂22 (6g,h)
            SL 2  2/3,1/3,z   P6/mmm (2c,d)
   3 b 2    SL 0  1/2,1/2,z   P6₂22 (3c,d)
   3 a 2    SL 2  0,0,z       P6/mmm (1a,b)
```

Space group   No.172   P6₄

Superlattices

```
SL  0  P(1,1,1)        density  1  add.gen.
SL  2  P(1,1,1/3)      density  3  add.gen.  0,0,1/3
```

Wyckoff letter   non-characteristic crystallographic orbits

```
  6 c 1     SL 0   x,-x,z       P6₄22 (6i,j)     x,0,z       P6₄22 (6g,h)
            SL 2   2/3,1/3,z    P6/mmm (2c,d)
  3 b 2     SL 0   1/2,1/2,z    P6₄22 (3c,d)
  3 a 2     SL 2   0,0,z        P6/mmm (1a,b)
```

Space group   No.173   P6₃

Superlattices

```
SL  0  P(1,1,1)        density  1  add.gen.
SL  1  P(1,1,1/2)      density  2  add.gen.  0,0,1/2
SL  3  P(1*,1*,1)      density  3  add.gen.  2/3,1/3,0
```

Wyckoff letter   non-characteristic crystallographic orbits

```
  6 c 1     SL 0   x,y,z        P6₃/m (6h)       x,-x,z      P6₃/mmc (6h)
                   x,0,z        P6₃/mcm (6g)
            SL 1   1/2,0,z      P6/mmm (3f,g)
            SL 3   1/3,1/3,z    P6₃/mmc (2c,d)
  2 b 3     SL 0   1/3,2/3,z    P6₃/mmc (2c,d)
  2 a 3     SL 1   0,0,z        P6/mmm (1a,b)
```

Space group No.174 P6̄

Superlattices

```
SL  0  P(1,1,1)       density  1  add.gen.
SL  1  P(1,1,1/2)     density  2  add.gen.  0,0,1/2
SL  3  P(1*,1*,1)     density  3  add.gen.  2/3,1/3,0
SL  7  P(1*,1*,1/2)   density  6  add.gen.  0,0,1/2  2/3,1/3,0
```

Wyckoff letter   non-characteristic crystallographic orbits

```
6 1 1    SL 0  x,-x,z        P6̄m2 (6n)    x,0,z          P6̄2m (6i)
               x,1/3,z       P6̄2m (6i)    x,2/3,z        P6̄2m (6i)
               1/2,0,z       P6/mmm (6i)  5/6,2/3,z      P6/mmm (6i)
               1/6,1/3,z     P6/mmm (6i)
         SL 1  x,y,1/4       P6̄ (3j,k)    x,-x,1/4       P6̄m2 (3j,k)
               x,0,1/4       P6̄2m (3f,g)  x,1/3,1/4      P6̄2m (3f,g)
               x,2/3,1/4     P6̄2m (3f,g)  1/2,0,1/4      P6/mmm (3f,g)
               5/6,2/3,1/4   P6/mmm (3f,g) 1/6,1/3,1/4   P6/mmm (3f,g)
         SL 3  1/3,1/3,z     P6/mmm (2e)  2/3,2/3,z      P6/mmm (2e)
         SL 7  1/3,1/3,1/4   P6/mmm (1a,b) 2/3,2/3,1/4   P6/mmm (1a,b)
3 k m    SL 0  x,-x,1/2      P6̄m2 (3j,k)  x,0,1/2        P6̄2m (3f,g)
               x,1/3,1/2     P6̄2m (3f,g)  x,2/3,1/2      P6̄2m (3f,g)
               1/2,0,1/2     P6/mmm (3f,g) 5/6,2/3,1/2   P6/mmm (3f,g)
               1/6,1/3,1/2   P6/mmm (3f,g)
         SL 3  1/3,1/3,1/2.  P6/mmm (1a,b) 2/3,2/3,1/2   P6/mmm (1a,b)
3 j m    SL 0  x,-x,0        P6̄m2 (3j,k)  x,0,0          P6̄2m (3f,g)
               x,1/3,0       P6̄2m (3f,g)  x,2/3,0        P6̄2m (3f,g)
               1/2,0,0       P6/mmm (3f,g) 5/6,2/3,0     P6/mmm (3f,g)
               1/6,1/3,0     P6/mmm (3f,g)
         SL 3  1/3,1/3,0     P6/mmm (1a,b) 2/3,2/3,0     P6/mmm (1a,b)
2 i 3    SL 0  2/3,1/3,z     P6/mmm (2e)
         SL 1  2/3,1/3,1/4   P6/mmm (1a,b)
2 h 3    SL 0  1/3,2/3,z     P6/mmm (2e)
         SL 1  1/3,2/3,1/4   P6/mmm (1a,b)
2 g 3    SL 0  0,0,z         P6/mmm (2e)
         SL 1  0,0,1/4       P6/mmm (1a,b)
1 f 6̄    SL 0  2/3,1/3,1/2   P6/mmm (1a,b)
1 e 6̄    SL 0  2/3,1/3,0     P6/mmm (1a,b)
1 d 6̄    SL 0  1/3,2/3,1/2   P6/mmm (1a,b)
1 c 6̄    SL 0  1/3,2/3,0     P6/mmm (1a,b)
1 b 6̄    SL 0  0,0,1/2       P6/mmm (1a,b)
1 a 6̄    SL 0  0,0,0         P6/mmm (1a,b)
```

Space group   No.175   P6/m

Superlattices

```
SL  0   P(1,1,1)          density  1  add.gen.
SL  1   P(1,1,1/2)        density  2  add.gen.  0,0,1/2
SL  3   P(1*,1*,1)        density  3  add.gen.  2/3,1/3,0
SL  7   P(1*,1*,1/2)      density  6  add.gen.  0,0,1/2  2/3,1/3,0
```

Wyckoff letter   non-characteristic crystallographic orbits

```
12 1 1      SL 0   x,-x,z          P6/mmm (12o)    x,0,z       P6/mmm (12n)
            SL 1   x,y,1/4         P6/m (6j,k)     x,-x,1/4    P6/mmm (61,m)
                   x,0,1/4         P6/mmm (6j,k)
            SL 3   1/3,1/3,z       P6/mmm (4h)
            SL 7   1/3,0,1/4       P6/mmm (2c,d)
 6 k m      SL 0   x,-x,1/2        P6/mmm (61,m)   x,0,1/2     P6/mmm (6j,k)
            SL 3   1/3,1/3,1/2     P6/mmm (2c,d)
 6 j m      SL 0   x,-x,0          P6/mmm (61,m)   x,0,0       P6/mmm (6j,k)
            SL 3   1/3,1/3,0       P6/mmm (2c,d)
 6 i 2      SL 0   1/2,0,z         P6/mmm (6i)
            SL 1   1/2,0,1/4       P6/mmm (3f,g)
 4 h 3      SL 0   1/3,2/3,z       P6/mmm (4h)
            SL 1   1/3,2/3,1/4     P6/mmm (2c,d)
 3 g 2/m    SL 0   1/2,0,1/2       P6/mmm (3f,g)
 3 f 2/m    SL 0   1/2,0,0         P6/mmm (3f,g)
 2 e 6      SL 0   0,0,z           P6/mmm (2e)
            SL 1   0,0,1/4         P6/mmm (1a,b)
 2 d 6̄      SL 0   1/3,2/3,1/2     P6/mmm (2c,d)
 2 c 6̄      SL 0   1/3,2/3,0       P6/mmm (2c,d)
 1 b 6/m    SL 0   0,0,1/2         P6/mmm (1a,b)
 1 a 6/m    SL 0   0,0,0           P6/mmm (1a,b)
```

Space group   No.176   P6$_3$/m

Superlattices

```
SL  0  P(1,1,1)        density  1  add.gen.
SL  1  P(1,1,1/2)      density  2  add.gen.  0,0,1/2
SL  3  P(1*,1*,1)      density  3  add.gen.  2/3,1/3,0
SL  4  P(1,1,1/4)      density  4  add.gen.  0,0,1/4
SL  7  P(1*,1*,1/2)    density  6  add.gen.  0,0,1/2  2/3,1/3,0
```

Wyckoff letter   non-characteristic crystallographic orbits

12 i 1	SL 0	x,-x,z	P6$_3$/mmc (12k)	x,0,z	P6$_3$/mcm (12k)
	SL 1	x,y,0	P6/m (6j,k)	x,-x,0	P6/mmm (6l,m)
		x,0,0	P6/mmm (6j,k)	1/2,0,z	P6/mmm (6i)
	SL 3	1/3,1/3,z	P6$_3$/mmc (4f)		
	SL 4	1/2,0,1/8	P6/mmm (3f,g)		
	SL 7	1/3,1/3,0	P6/mmm (2c,d)		
6 h m	SL 0	x,-x,1/4	P6$_3$/mmc (6h)	x,0,1/4	P6$_3$/mcm (6g)
	SL 1	1/2,0,1/4	P6/mmm (3f,g)		
	SL 3	1/3,1/3,1/4	P6$_3$/mmc (2c,d)	2/3,2/3,1/4	P6$_3$/mmc (2c,d)
6 g $\bar{1}$	SL 1	1/2,0,0	P6/mmm (3f,g)		
4 f 3	SL 0	1/3,2/3,z	P6$_3$/mmc (4f)		
	SL 1	1/3,2/3,0	P6/mmm (2c,d)		
4 e 3	SL 1	0,0,z	P6/mmm (2e)		
	SL 4	0,0,1/8	P6/mmm (1a,b)		
2 d $\bar{6}$	SL 0	2/3,1/3,1/4	P6$_3$/mmc (2c,d)		
2 c $\bar{6}$	SL 0	1/3,2/3,1/4	P6$_3$/mmc (2c,d)		
2 b $\bar{3}$	SL 1	0,0,0	P6/mmm (1a,b)		
2 a $\bar{6}$	SL 1	0,0,1/4	P6/mmm (1a,b)		

Space group No.177 P622

Superlattices

SL	0	P(1,1,1)	density	1	add.gen.	
SL	1	P(1,1,1/2)	density	2	add.gen.	0,0,1/2
SL	3	P(1*,1*,1)	density	3	add.gen.	2/3,1/3,0
SL	7	P(1*,1*,1/2)	density	6	add.gen.	0,0,1/2 2/3,1/3,0

Wyckoff letter   non-characteristic crystallographic orbits

12 n 1	SL 0	x,y,1/4	P6/mcc (12l)	x,y,1/2	P6/mmm (12p,q)
		x,y,0	P6/mmm (12p,q)	x,-x,z	P6/mmm (12o)
		x,0,z	P6/mmm (12n)		
	SL 1	x,0,1/4	P6/mmm (6j,k)	x,-x,1/4	P6/mmm (6l,m)
	SL 3	1/3,1/3,z	P6/mmm (4h)		
	SL 7	1/3,1/3,1/4	P6/mmm (2c,d)		
6 m 2	SL 0	x,-x,1/2	P6/mmm (6l,m)		
6 l 2	SL 0	x,-x,0	P6/mmm (6l,m)		
6 k 2	SL 0	x,0,1/2	P6/mmm (6j,k)		
	SL 3	1/3,0,1/2	P6/mmm (2c,d)		
6 j 2	SL 0	x,0,0	P6/mmm (6j,k)		
	SL 3	1/3,0,0	P6/mmm (2c,d)		
6 i 2	SL 0	1/2,0,z	P6/mmm (6i)		
	SL 1	1/2,0,1/4	P6/mmm (3f,g)		
4 h 3	SL 0	1/3,2/3,z	P6/mmm (4h)		
	SL 1	1/3,2/3,1/4	P6/mmm (2c,d)		
3 g 222	SL 0	1/2,0,1/2	P6/mmm (3f,g)		
3 f 222	SL 0	1/2,0,0	P6/mmm (3f,g)		
2 e 6	SL 0	0,0,z	P6/mmm (2e)		
	SL 1	0,0,1/4	P6/mmm (1a,b)		
2 d 32	SL 0	1/3,2/3,1/2	P6/mmm (2c,d)		
2 c 32	SL 0	1/3,2/3,0	P6/mmm (2c,d)		
1 b 622	SL 0	0,0,1/2	P6/mmm (1a,b)		
1 a 622	SL 0	0,0,0	P6/mmm (1a,b)		

Space group No.178 P6$_1$22

Superlattices

```
SL  1   P(1,1,1/2)        density  2  add.gen.  0,0,1/2
SL  2   P(1,1,1/3)        density  3  add.gen.  0,0,1/3
SL  4   P(1,1,1/4)        density  4  add.gen.  0,0,1/4
SL  6   P(1,1,1/6)        density  6  add.gen.  0,0,1/6
SL 10   P(1,1,1/12)       density 12  add.gen.  0,0,1/12
```

Wyckoff letter   non-characteristic crystallographic orbits

```
 12 c 1     SL 1   1/2,0,z      P6₂22 (6f)       x,0,1/4       P6₂22 (6g,h)
                   x,2x,0       P6₂22 (6i,j)
            SL 2   2/3,1/3,z    P6₃/mmc (4f)
            SL 4   1/2,0,1/8    P6₄22 (3c,d)
            SL 6   0,0,z        P6/mmm (2e)      2/3,1/3,0     P6/mmm (2c,d)
            SL10   0,0,1/24     P6/mmm (1a,b)
  6 b 2     SL 1   1/2,0,1/4    P6₂22 (3c,d)
            SL 2   1/3,2/3,1/4  P6₃/mmc (2c,d)   2/3,1/3,1/4   P6₃/mmc (2c,d)
            SL 6   0,0,1/4      P6/mmm (1a,b)
  6 a 2     SL 1   1/2,0,0      P6₂22 (3c,d)
            SL 6   0,0,0        P6/mmm (1a,b)
```

Space group No.179 P6$_5$22

Superlattices

```
SL  1   P(1,1,1/2)        density  2  add.gen.  0,0,1/2
SL  2   P(1,1,1/3)        density  3  add.gen.  0,0,1/3
SL  4   P(1,1,1/4)        density  4  add.gen.  0,0,1/4
SL  6   P(1,1,1/6)        density  6  add.gen.  0,0,1/6
SL 10   P(1,1,1/12)       density 12  add.gen.  0,0,1/12
```

Wyckoff letter   non-characteristic crystallographic orbits

```
 12 c 1     SL 1   1/2,0,z      P6₄22 (6f)       x,0,1/4       P6₄22 (6g,h)
                   x,2x,0       P6₄22 (6i,j)
            SL 2   2/3,1/3,z    P6₃/mmc (4f)
            SL 4   1/2,0,1/8    P6₂22 (3c,d)
            SL 6   0,0,z        P6/mmm (2e)      2/3,1/3,0     P6/mmm (2c,d)
            SL10   0,0,1/24     P6/mmm (1a,b)
  6 b 2     SL 1   1/2,0,3/4    P6₄22 (3c,d)
            SL 2   2/3,1/3,3/4  P6₃/mmc (2c,d)   1/3,2/3,3/4   P6₃/mmc (2c,d)
            SL 6   0,0,3/4      P6/mmm (1a,b)
  6 a 2     SL 1   1/2,0,0      P6₄22 (3c,d)
            SL 6   0,0,0        P6/mmm (1a,b)
```

Space group No.180 P6₂22

Superlattices

```
SL  1  P(1,1,1/2)       density  2  add.gen.  0,0,1/2
SL  2  P(1,1,1/3)       density  3  add.gen.  0,0,1/3
SL  6  P(1,1,1/6)       density  6  add.gen.  0,0,1/6
```

Wyckoff letter   non-characteristic crystallographic orbits

```
 12 k 1      SL 1   x,0,1/4        P6₄22 (6g,h)    x,2x,1/4    P6₄22 (6i,j)
             SL 2   1/3,2/3,z      P6/mmm (4h)
             SL 6   1/3,2/3,1/4    P6/mmm (2c,d)
  6 j 2      SL 2   1/3,2/3,1/2    P6/mmm (2c,d)
  6 i 2      SL 2   1/3,2/3,0      P6/mmm (2c,d)
  6 f 2      SL 1   1/2,0,1/4      P6₄22 (3c,d)
  6 e 2      SL 2   0,0,z          P6/mmm (2e)
             SL 6   0,0,1/12       P6/mmm (1a,b)
  3 b 222    SL 2   0,0,1/2        P6/mmm (1a,b)
  3 a 222    SL 2   0,0,0          P6/mmm (1a,b)
```


Space group No.181 P6₄22

Superlattices

```
SL  1  P(1,1,1/2)       density  2  add.gen.  0,0,1/2
SL  2  P(1,1,1/3)       density  3  add.gen.  0,0,1/3
SL  6  P(1,1,1/6)       density  6  add.gen.  0,0,1/6
```

Wyckoff letter   non-characteristic crystallographic orbits

```
 12 k 1      SL 1   x,0,1/4        P6₂22 (6g,h)    x,2x,1/4    P6₂22 (6i,j)
             SL 2   1/3,2/3,z      P6/mmm (4h)
             SL 6   1/3,2/3,1/4    P6/mmm (2c,d)
  6 j 2      SL 2   1/3,2/3,1/2    P6/mmm (2c,d)
  6 i 2      SL 2   1/3,2/3,0      P6/mmm (2c,d)
  6 f 2      SL 1   1/2,0,1/4      P6₂22 (3c,d)
  6 e 2      SL 2   0,0,z          P6/mmm (2e)
             SL 6   0,0,1/12       P6/mmm (1a,b)
  3 b 222    SL 2   0,0,1/2        P6/mmm (1a,b)
  3 a 222    SL 2   0,0,0          P6/mmm (1a,b)
```

Space group No.182 P6₃22

Superlattices

```
SL  0  P(1,1,1)      density  1  add.gen.
SL  1  P(1,1,1/2)    density  2  add.gen.  0,0,1/2
SL  3  P(1*,1*,1)    density  3  add.gen.  2/3,1/3,0
SL  4  P(1,1,1/4)    density  4  add.gen.  0,0,1/4
SL  7  P(1*,1*,1/2)  density  6  add.gen.  0,0,1/2  2/3,1/3,0
```

Wyckoff letter   non-characteristic crystallographic orbits

```
12 i 1    SL 0   x,y,0        P6₃/mcm (12j)   x,0,z      P6₃/mcm (12k)
                 x,2x,z       P6₃/mmc (12k)   x,y,1/4    P6₃/mmc (12j)
          SL 1   x,0,1/4      P6/mmm (6j,k)   1/2,0,z    P6/mmm (6i)
                 x,2x,0       P6/mmm (6l,m)
          SL 3   1/3,1/3,z    P6₃/mmc (4f)
          SL 4   1/2,0,1/8    P6/mmm (3f,g)
          SL 7   1/3,1/3,1/4  P6/mmm (2c,d)
 6 h 2    SL 0   x,2x,1/4     P6₃/mmc (6h)
          SL 1   1/2,0,1/4    P6/mmm (3f,g)
 6 g 2    SL 0   x,0,0        P6₃/mcm (6g)
          SL 1   1/2,0,0      P6/mmm (3f,g)
          SL 3   1/3,0,0      P6₃/mmc (2c,d)   2/3,0,0    P6₃/mmc (2c,d)
 4 f 3    SL 0   1/3,2/3,z    P6₃/mmc (4f)
          SL 1   1/3,2/3,0    P6/mmm (2c,d)
 4 e 3    SL 1   0,0,z        P6/mmm (2e)
          SL 4   0,0,1/8      P6/mmm (1a,b)
 2 d 32   SL 0   1/3,2/3,3/4  P6₃/mmc (2c,d)
 2 c 32   SL 0   1/3,2/3,1/4  P6₃/mmc (2c,d)
 2 b 32   SL 1   0,0,1/4      P6/mmm (1a,b)
 2 a 32   SL 1   0,0,0        P6/mmm (1a,b)
```

Space group   No.183   P6mm

Superlattices

```
SL  0  P(1,1,1)        density  1  add.gen.
SL  3  P(1*,1*,1)      density  3  add.gen.  2/3,1/3,0
```

Wyckoff letter   non-characteristic crystallographic orbits

```
  12 f 1      SL 0   x,y,z      P6/mmm (12p,q)
   6 e m      SL 0   x,-x,z     P6/mmm (6l,m)
   6 d m      SL 0   x,0,z      P6/mmm (6j,k)
              SL 3   1/3,0,z    P6/mmm (2c,d)
   3 c mm2    SL 0   1/2,0,z    P6/mmm (3f,g)
   2 b 3m     SL 0   1/3,2/3,z  P6/mmm (2c,d)
   1 a 6mm    SL 0   0,0,z      P6/mmm (1a,b)
```

Space group   No.184   P6cc

Superlattices

```
SL  0  P(1,1,1)        density  1  add.gen.
SL  1  P(1,1,1/2)      density  2  add.gen.  0,0,1/2
SL  7  P(1*,1*,1/2)    density  6  add.gen.  0,0,1/2  2/3,1/3,0
```

Wyckoff letter   non-characteristic crystallographic orbits

```
  12 d 1      SL 0   x,y,z      P6/mcc (12l)
              SL 1   x,0,z      P6/mmm (6j,k)   x,-x,z     P6/mmm (6l,m)
              SL 7   1/3,1/3,z  P6/mmm (2c,d)
   6 c 2      SL 1   1/2,0,z    P6/mmm (3f,g)
   4 b 3      SL 1   1/3,2/3,z  P6/mmm (2c,d)
   2 a 6      SL 1   0,0,z      P6/mmm (1a,b)
```

Space group No.185 P6₃cm

Superlattices

```
SL  0  P(1,1,1)        density  1  add.gen.
SL  1  P(1,1,1/2)      density  2  add.gen.   0,0,1/2
SL  3  P(1*,1*,1)      density  3  add.gen.   2/3,1/3,0
```

Wyckoff letter   non-characteristic crystallographic orbits

```
  12 d 1      SL 0   x,y,z       P6₃/mcm (12j)
              SL 1   x,-x,z      P6/mmm (6l,m)
   6 c m      SL 0   x,0,z       P6₃/mcm (6g)
              SL 1   1/2,0,z     P6/mmm (3f,g)
              SL 3   2/3,0,z     P6₃/mmc (2c,d)
   4 b 3      SL 1   1/3,2/3,z   P6/mmm (2c,d)
   2 a 3m     SL 1   0,0,z       P6/mmm (1a,b)
```

Space group No.186 P6₃mc

Superlattices

```
SL  0  P(1,1,1)        density  1  add.gen.
SL  1  P(1,1,1/2)      density  2  add.gen.   0,0,1/2
SL  7  P(1*,1*,1/2)    density  6  add.gen.   0,0,1/2  2/3,1/3,0
```

Wyckoff letter   non-characteristic crystallographic orbits

```
  12 d 1      SL 0   x,y,z       P6₃/mmc (12j)
              SL 1   x,0,z       P6/mmm (6j,k)
              SL 7   1/3,1/3,z   P6/mmm (2c,d)
   6 c m      SL 0   x,-x,z      P6₃/mmc (6h)
              SL 1   1/2,1/2,z   P6/mmm (3f,g)
   2 b 3m     SL 0   1/3,2/3,z   P6₃/mmc (2c,d)
   2 a 3m     SL 1   0,0,z       P6/mmm (1a,b)
```

Space group No.187 P6̄m2

Superlattices

SL	0	P(1,1,1)	density	1	add.gen.
SL	1	P(1,1,1/2)	density	2	add.gen.	,0,0,1/2
SL	3	P(1*,1*,1)	density	3	add.gen.	2/3,1/3,0
SL	7	P(1*,1*,1/2)	density	6	add.gen.	0,0,1/2 2/3,1/3,0

Wyckoff letter   non-characteristic crystallographic orbits

12 o 1	SL 0	x,0,z	P6/mmm (12n)	x,1/3,z	P6/mmm (12n)
		x,2/3,z	P6/mmm (12n)		
	SL 1	x,y,1/4	P6̄m2 (6l,m)	x,0,1/4	P6/mmm (6j,k)
	SL 3	1/3,1/3,z	P6/mmm (4h)		
	SL 7	1/3,1/3,1/4	P6/mmm (2c,d)		
6 n m	SL 0	1/2,1/2,z	P6/mmm (6i)	5/6,1/6,z	P6/mmm (6i)
		1/6,5/6,z	P6/mmm (6i)		
	SL 1	x,-x,1/4	P6̄m2 (3j,k)	1/2,1/2,1/4	P6/mmm (3f,g)
		5/6,1/6,1/4	P6/mmm (3f,g)	1/6,5/6,1/4	P6/mmm (3f,g)
6 m m	SL 0	x,0,1/2	P6/mmm (6j,k)	x,1/3,1/2	P6/mmm (6j,k)
		x,2/3,1/2	P6/mmm (6j,k)		
	SL 3	1/3,1/3,1/2	P6/mmm (2c,d)		
6 l m	SL 0	x,0,0	P6/mmm (6j,k)	x,1/3,0	P6/mmm (6j,k)
		x,2/3,0	P6/mmm (6j,k)		
	SL 3	1/3,1/3,0	P6/mmm (2c,d)		
3 k mm2	SL 0	1/2,1/2,1/2	P6/mmm (3f,g)	5/6,1/6,1/2	P6/mmm (3f,g)
		1/6,5/6,1/2	P6/mmm (3f,g)		
3 j mm2	SL 0	1/2,1/2,0	P6/mmm (3f,g)	5/6,1/6,0	P6/mmm (3f,g)
		1/6,5/6,0	P6/mmm (3f,g)		
2 i 3m	SL 0	2/3,1/3,z	P6/mmm (2e)		
	SL 1	2/3,1/3,1/4	P6/mmm (1a,b)		
2 h 3m	SL 0	1/3,2/3,z	P6/mmm (2e)		
	SL 1	1/3,2/3,1/4	P6/mmm (1a,b)		
2 g 3m	SL 0	0,0,z	P6/mmm (2e)		
	SL 1	0,0,1/4	P6/mmm (1a,b)		
1 f 6̄m2	SL 0	2/3,1/3,1/2	P6/mmm (1a,b)		
1 e 6̄m2	SL 0	2/3,1/3,0	P6/mmm (1a,b)		
1 d 6̄m2	SL 0	1/3,2/3,1/2	P6/mmm (1a,b)		
1 c 6̄m2	SL 0	1/3,2/3,0	P6/mmm (1a,b)		
1 b 6̄m2	SL 0	0,0,1/2	P6/mmm (1a,b)		
1 a 6̄m2	SL 0	0,0,0	P6/mmm (1a,b)		

Space group   No.188   P$\bar{6}$c2

Superlattices

SL	0	P(1,1,1)	density	1	add.gen.		
SL	1	P(1,1,1/2)	density	2	add.gen.	0,0,1/2	
SL	3	P(1*,1*,1)	density	3	add.gen.	2/3,1/3,0	
SL	4	P(1,1,1/4)	density	4	add.gen.	0,0,1/4	
SL	7	P(1*,1*,1/2)	density	6	add.gen.	0,0,1/2	2/3,1/3,0

Wyckoff letter   non-characteristic crystallographic orbits

12 1 1	SL 0	x,0,z	P6$_3$/mcm (12k)	x,1/3,z	P6$_3$/mcm (12k)
		x,2/3,z	P6$_3$/mcm (12k)		
	SL 1	x,-x,z	P$\bar{6}$m2 (6n)	x,y,0	P$\bar{6}$m2 (6l,m)
		1/2,0,z	P6/mmm (6i)	5/6,2/3,z	P6/mmm (6i)
		1/6,1/3,z	P6/mmm (6i)	x,0,0	P6/mmm (6j,k)
		x,1/3,0	P6/mmm (6j,k)	x,2/3,0	P6/mmm (6j,k)
	SL 3	1/3,1/3,z	P6$_3$/mmc (4f)		
	SL 4	x,-x,1/8	P$\bar{6}$m2 (3j,k)	1/2,0,1/8	P6/mmm (3f,g)
		5/6,2/3,1/8	P6/mmm (3f,g)	1/6,1/3,1/8	P6/mmm (3f,g)
	SL 7	1/3,1/3,0	P6/mmm (2c,d)		
6 k m	SL 0	x,0,1/4	P6$_3$/mcm (6g)	x,1/3,1/4	P6$_3$/mcm (6g)
		x,2/3,1/4	P6$_3$/mcm (6g)		
	SL 1	x,-x,1/4	P$\bar{6}$m2 (3j,k)	1/2,0,1/4	P6/mmm (3f,g)
		5/6,2/3,1/4	P6/mmm (3f,g)	1/6,1/3,1/4	P6/mmm (3f,g)
	SL 3	1/3,1/3,1/4	P6$_3$/mmc (2c,d)	2/3,2/3,1/4	P6$_3$/mmc (2c,d)
6 j 2	SL 1	x,-x,0	P$\bar{6}$m2 (3j,k)	1/2,1/2,0	P6/mmm (3f,g)
		5/6,1/6,0	P6/mmm (3f,g)	1/6,5/6,0	P6/mmm (3f,g)
4 i 3	SL 1	2/3,1/3,z	P6/mmm (2e)		
	SL 4	2/3,1/3,1/8	P6/mmm (1a,b)		
4 h 3	SL 1	1/3,2/3,z	P6/mmm (2e)		
	SL 4	1/3,2/3,1/8	P6/mmm (1a,b)		
4 g 3	SL 1	0,0,z	P6/mmm (2e)		
	SL 4	0,0,1/8	P6/mmm (1a,b)		
2 f $\bar{6}$	SL 1	2/3,1/3,1/4	P6/mmm (1a,b)		
2 e 32	SL 1	2/3,1/3,0	P6/mmm (1a,b)		
2 d $\bar{6}$	SL 1	1/3,2/3,1/4	P6/mmm (1a,b)		
2 c 32	SL 1	1/3,2/3,0	P6/mmm (1a,b)		
2 b $\bar{6}$	SL 1	0,0,1/4	P6/mmm (1a,b)		
2 a 32	SL 1	0,0,0	P6/mmm (1a,b)		

Space group   No.189   P6̄2m

Superlattices

```
SL  0  P(1,1,1)        density  1  add.gen.
SL  1  P(1,1,1/2)      density  2  add.gen.  0,0,1/2
SL  3  P(1*,1*,1)      density  3  add.gen.  2/3,1/3,0
SL  7  P(1*,1*,1/2)    density  6  add.gen.  0,0,1/2  2/3,1/3,0
```

Wyckoff letter   non-characteristic crystallographic orbits

```
12 l 1      SL 0  x,2x,z       P6/mmm (12o)
            SL 1  x,y,1/4      P6̄2m (6j,k)      x,2x,1/4      P6/mmm (61,m)
 6 k m      SL 0  x,2x,1/2     P6/mmm (61,m)
 6 j m      SL 0  x,2x,0       P6/mmm (61,m)
 6 i m      SL 0  1/2,0,z      P6/mmm (6i)
            SL 1  x,0,1/4      P6̄2m (3f,g)      1/2,0,1/4     P6/mmm (3f,g)
            SL 3  1/3,0,z      P6/mmm (2e)      2/3,0,z       P6/mmm (2e)
            SL 7  1/3,0,1/4    P6/mmm (1a,b)    2/3,0,1/4     P6/mmm (1a,b)
 4 h 3      SL 0  1/3,2/3,z    P6/mmm (4h)
            SL 1  1/3,2/3,1/4  P6/mmm (2c,d)
 3 g mm2    SL 0  1/2,0,1/2    P6/mmm (3f,g)
            SL 3  1/3,0,1/2    P6/mmm (1a,b)    2/3,0,1/2     P6/mmm (1a,b)
 3 f mm2    SL 0  1/2,0,0      P6/mmm (3f,g)
            SL 3  1/3,0,0      P6/mmm (1a,b)    2/3,0,0       P6/mmm (1a,b)
 2 e 3m     SL 0  0,0,z        P6/mmm (2e)
            SL 1  0,0,1/4      P6/mmm (1a,b)
 2 d 6̄      SL 0  1/3,2/3,1/2  P6/mmm (2c,d)
 2 c 6̄      SL 0  1/3,2/3,0    P6/mmm (2c,d)
 1 b 6̄2m    SL 0  0,0,1/2      P6/mmm (1a,b)
 1 a 6̄2m    SL 0  0,0,0        P6/mmm (1a,b)
```

Space group   No.190   P6̄2c

Superlattices

```
SL  0   P(1,1,1)        density  1  add.gen.
SL  1   P(1,1,1/2)      density  2  add.gen.  0,0,1/2
SL  4   P(1,1,1/4)      density  4  add.gen.  0,0,1/4
SL  7   P(1*,1*,1/2)    density  6  add.gen.  0,0,1/2  2/3,1/3,0
SL 11   P(1*,1*,1/4)    density 12  add.gen.  0,0,1/4  2/3,1/3,0
```

Wyckoff letter   non-characteristic crystallographic orbits

12 i 1	SL 0	x,2x,z	P6₃/mmc (12k)	
	SL 1	x,0,z	P6̄2m (6i)	x,y,0	P6̄2m (6j,k)
		1/2,0,z	P6/mmm (6i)	x,-x,0	P6/mmm (6l,m)
	SL 4	x,0,1/8	P6̄2m (3f,g)	1/2,0,1/8	P6/mmm (3f,g)
	SL 7	2/3,2/3,z	P6/mmm (2e)	1/3,1/3,z	P6/mmm (2e)
	SL11	2/3,2/3,1/8	P6/mmm (1a,b)	1/3,1/3,1/8	P6/mmm (1a,b)
6 h m	SL 0	x,2x,1/4	P6₃/mmc (6h)	
	SL 1	x,0,1/4	P6̄2m (3f,g)	1/2,0,1/4	P6/mmm (3f,g)
	SL 7	2/3,2/3,1/4	P6/mmm (1a,b)	1/3,1/3,1/4	P6/mmm (1a,b)
6 g 2	SL 1	x,0,0	P6̄2m (3f,g)	1/2,0,0	P6/mmm (3f,g)
	SL 7	1/3,0,0	P6/mmm (1a,b)	2/3,0,0	P6/mmm (1a,b)
4 f 3	SL 0	1/3,2/3,z	P6₃/mmc (4f)	
	SL 1	1/3,2/3,0	P6/mmm (2c,d)	
4 e 3	SL 1	0,0,z	P6/mmm (2e)	
	SL 4	0,0,1/8	P6/mmm (1a,b)	
2 d 6̄	SL 0	2/3,1/3,1/4	P6₃/mmc (2c,d)	
2 c 6̄	SL 0	1/3,2/3,1/4	P6₃/mmc (2c,d)	
2 b 6̄	SL 1	0,0,1/4	P6/mmm (1a,b)	
2 a 32	SL 1	0,0,0	P6/mmm (1a,b)	

Space group   No.191   P6/mmm

Superlattices

```
SL  1   P(1,1,1/2)      density  2  add.gen.  0,0,1/2
SL  3   P(1*,1*,1)      density  3  add.gen.  2/3,1/3,0
SL  7   P(1*,1*,1/2)    density  6  add.gen.  0,0,1/2  2/3,1/3,0
```

Wyckoff letter   non-characteristic crystallographic orbits

24 r 1	SL 1	x,y,1/4	P6/mmm (12p,q)
12 o m	SL 1	x,2x,1/4	P6/mmm (6l,m)
12 n m	SL 1	x,0,1/4	P6/mmm (6j,k)
	SL 3	1/3,0,z	P6/mmm (4h)
	SL 7	1/3,0,1/4	P6/mmm (2c,d)
6 k mm2	SL 3	1/3,0,1/2	P6/mmm (2c,d)
6 j mm2	SL 3	1/3,0,0	P6/mmm (2c,d)
6 i mm2	SL 1	1/2,0,1/4	P6/mmm (3f,g)
4 h 3m	SL 1	1/3,2/3,1/4	P6/mmm (2c,d)
2 e 6mm	SL 1	0,0,1/4	P6/mmm (1a,b)

Space group   No.192   P6/mcc

Superlattices

SL	1	P(1,1,1/2)	density	2	add.gen.	0,0,1/2	
SL	4	P(1,1,1/4)	density	4	add.gen.	0,0,1/4	
SL	7	P(1*,1*,1/2)	density	6	add.gen.	0,0,1/2	2/3,1/3,0
SL	11	P(1*,1*,1/4)	density	12	add.gen.	0,0,1/4	2/3,1/3,0

Wyckoff letter   non-characteristic crystallographic orbits

24 m 1	SL 1	x,y,1/4	P6/mmm (12p,q)	x,0,z	P6/mmm (12n)
		x,2x,z	P6/mmm (12o)		
	SL 4	x,2x,1/8	P6/mmm (6l,m)	x,0,1/8	P6/mmm (6j,k)
	SL 7	1/3,1/3,z	P6/mmm (4h)		
	SL11	1/3,1/3,1/8	P6/mmm (2c,d)		
12 l m	SL 1	x,0,0	P6/mmm (6j,k)	x,2x,0	P6/mmm (6l,m)
	SL 7	1/3,1/3,0	P6/mmm (2c,d)		
12 k 2	SL 1	x,2x,1/4	P6/mmm (6l,m)		
12 j 2	SL 1	x,0,1/4	P6/mmm (6j,k)		
	SL 7	1/3,0,1/4	P6/mmm (2c,d)		
12 i 2	SL 1	1/2,0,z	P6/mmm (6i)		
	SL 4	1/2,0,1/8	P6/mmm (3f,g)		
8 h 3	SL 1	1/3,2/3,z	P6/mmm (4h)		
	SL 4	1/3,2/3,1/8	P6/mmm (2c,d)		
6 g 2/m	SL 1	1/2,0,0	P6/mmm (3f,g)		
6 f 222	SL 1	1/2,0,1/4	P6/mmm (3f,g)		
4 e 6	SL 1	0,0,z	P6/mmm (2e)		
	SL 4	0,0,1/8	P6/mmm (1a,b)		
4 d $\bar{6}$	SL 1	1/3,2/3,0	P6/mmm (2c,d)		
4 c 32	SL 1	1/3,2/3,1/4	P6/mmm (2c,d)		
2 b 6/m	SL 1	0,0,0	P6/mmm (1a,b)		
2 a 622	SL 1	0,0,1/4	P6/mmm (1a,b)		

Space group No.193 P6$_3$/mcm

Superlattices

SL	1	P(1,1,1/2)	density	2	add.gen.	0,0,1/2	
SL	3	P(1*,1*,1)	density	3	add.gen.	2/3,1/3,0	
SL	4	P(1,1,1/4)	density	4	add.gen.	0,0,1/4	
SL	7	P(1*,1*,1/2)	density	6	add.gen.	0,0,1/2	2/3,1/3,0

Wyckoff letter		non-characteristic crystallographic orbits			
24 l 1	SL 1	x,y,0	P6/mmm (12p,q)	x,2x,z	P6/mmm (12o)
	SL 4	x,2x,1/8	P6/mmm (6l,m)		
12 k m	SL 1	x,0,0	P6/mmm (6j,k)	1/2,0,z	P6/mmm (6i)
	SL 3	1/3,0,z	P6/mmm (4h)		
	SL 4	1/2,0,1/8	P6/mmm (3f,g)		
	SL 7	1/3,0,0	P6/mmm (2c,d)		
12 j m	SL 1	x,2x,1/4	P6/mmm (6l,m)		
12 i 2	SL 1	x,2x,0	P6/mmm (6l,m)		
8 h 3	SL 1	1/3,2/3,z	P6/mmm (4h)		
	SL 4	1/3,2/3,1/8	P6/mmm (2c,d)		
6 g mm2	SL 1	1/2,0,1/4	P6/mmm (3f,g)		
	SL 3	2/3,0,1/4	P6$_3$/mmc (2c,d)	1/3,0,1/4	P6$_3$/mmc (2c,d)
6 f 2/m	SL 1	1/2,0,0	P6/mmm (3f,g)		
4 e 3m	SL 1	0,0,z	P6/mmm (2e)		
	SL 4	0,0,1/8	P6/mmm (1a,b)		
4 d 32	SL 1	1/3,2/3,0	P6/mmm (2c,d)		
4 c $\bar{6}$	SL 1	1/3,2/3,1/4	P6/mmm (2c,d)		
2 b $\bar{3}$m	SL 1	0,0,0	P6/mmm (1a,b)		
2 a $\bar{6}$2m	SL 1	0,0,1/4	P6/mmm (1a,b)		

Space group No.194 P6₃/mmc

Superlattices

```
SL  1  P(1,1,1/2)      density  2  add.gen.  0,0,1/2
SL  4  P(1,1,1/4)      density  4  add.gen.  0,0,1/4
SL  7  P(1*,1*,1/2)    density  6  add.gen.  0,0,1/2  2/3,1/3,0
SL 11  P(1*,1*,1/4)    density 12  add.gen.  0,0,1/4  2/3,1/3,0
```

Wyckoff letter   non-characteristic crystallographic orbits

```
24 l 1    SL 1  x,y,0        P6/mmm (12p,q)  x,0,z        P6/mmm (12n)
          SL 4  x,0,1/8      P6/mmm (6j,k)
          SL 7  1/3,1/3,z    P6/mmm (4h)
          SL11  1/3,1/3,1/8  P6/mmm (2c,d)
12 k m    SL 1  x,2x,0       P6/mmm (6l,m)   1/2,0,z      P6/mmm (6i)
          SL 4  1/2,0,1/8    P6/mmm (3f,g)
12 j m    SL 1  x,0,1/4      P6/mmm (6j,k)
          SL 7  1/3,1/3,1/4  P6/mmm (2c,d)
12 i 2    SL 1  x,0,0        P6/mmm (6j,k)
          SL 7  1/3,0,0      P6/mmm (2c,d)
 6 h mm2  SL 1  1/2,0,1/4    P6/mmm (3f,g)
 6 g 2/m  SL 1  1/2,0,0      P6/mmm (3f,g)
 4 f 3m   SL 1  1/3,2/3,0    P6/mmm (2c,d)
 4 e 3m   SL 1  0,0,z        P6/mmm (2e)
          SL 4  0,0,1/8      P6/mmm (1a,b)
 2 b 6̄m2  SL 1  0,0,1/4      P6/mmm (1a,b)
 2 a 3̄m   SL 1  0,0,0        P6/mmm (1a,b)
```

3.7. Cubic system

Space group No.195 P23

Superlattices

```
SL  0   P(1,1,1)           density  1  add.gen.
SL  2   F(1,1,1)           density  4  add.gen.   0,1/2,1/2   1/2,0,1/2
```

Wyckoff letter		non-characteristic crystallographic orbits			
12 j 1	SL 0	0,y,z	Pm$\overline{3}$ (12j,k)	1/2,y,z	Pm$\overline{3}$ (12j,k)
		x,x,z	P$\overline{4}$3m (12i)	1/4,x,1/2+x	P4$_2$32 (12k,l`
		1/4,x,1/2-x	P4$_2$32 (12k,l)	0,x,x	Pm$\overline{3}$m (12i,j)
		1/2,x,x	Pm$\overline{3}$m (12i,j)		
6 i 2	SL 0	x,1/2,1/2	Pm$\overline{3}$m (6e,f)		
6 h 2	SL 0	x,1/2,0	Pm$\overline{3}$ (6f,g)	1/4,1/2,0	Pm$\overline{3}$n (6c,d)
6 g 2	SL 0	x,0,1/2	Pm$\overline{3}$ (6f,g)	1/4,0,1/2	Pm$\overline{3}$n (6c,d)
6 f 2	SL 0	x,0,0	Pm$\overline{3}$m (6e,f)		
4 e 3	SL 0	x,x,x	P$\overline{4}$3m (4e)		
	SL 2	3/4,3/4,3/4	Fm$\overline{3}$m (4a,b)	1/4,1/4,1/4	Fm$\overline{3}$m (4a,b)
3 d 222	SL 0	1/2,0,0	Pm$\overline{3}$m (3c,d)		
3 c 222	SL 0	0,1/2,1/2	Pm$\overline{3}$m (3c,d)		
1 b 23	SL 0	1/2,1/2,1/2	Pm$\overline{3}$m (1a,b)		
1 a 23	SL 0	0,0,0	Pm$\overline{3}$m (1a,b)		

Space group No.196 F23

Superlattices

```
SL  0   F(1,1,1)              density  1  add.gen.
SL  1   P(1/2,1/2,1/2)        density  2  add.gen.   1/2,0,0
```

Wyckoff letter		non-characteristic crystallographic orbits			
48 h 1	SL 0	0,y,z	Fm$\overline{3}$ (48h)	1/4,y,z	Fm$\overline{3}$ (48h)
		x,x,z	F$\overline{4}$3m (48h)	1/8,x,1/4-x	F4$_1$32 (48g)
		1/8,x,1/4+x	F4$_1$32 (48g)	0,x,x	Fm$\overline{3}$m (48h,i)
		1/4,x,x	Fm$\overline{3}$m (48h,i)	1/2,x,x	Fm$\overline{3}$m (48h,i)
		3/4,x,x	Fm$\overline{3}$m (48h,i)		
	SL 1	1/4,y,0	Pm$\overline{3}$ (6f,g)	0,y,1/4	Pm$\overline{3}$ (6f,g)
		1/4,1/8,0	Pm$\overline{3}$n (6c,d)	0,1/8,1/4	Pm$\overline{3}$n (6c,d)
24 g 2	SL 0	x,1/4,1/4	Fm$\overline{3}$m (24e)		
	SL 1	0,1/4,1/4	Pm$\overline{3}$m (3c,d)		
24 f 2	SL 0	x,0,0	Fm$\overline{3}$m (24e)		
	SL 1	1/4,0,0	Pm$\overline{3}$m (3c,d)		
16 e 3	SL 0	x,x,x	F$\overline{4}$3m (16e)	1/8,1/8,1/8	Fd$\overline{3}$m (16c,d)
		3/8,3/8,3/8	Fd$\overline{3}$m (16c,d)	5/8,5/8,5/8	Fd$\overline{3}$m (16c,d)
		7/8,7/8,7/8	Fd$\overline{3}$m (16c,d)		
4 d 23	SL 0	3/4,3/4,3/4	Fm$\overline{3}$m (4a,b)		
4 c 23	SL 0	1/4,1/4,1/4	Fm$\overline{3}$m (4a,b)		
4 b 23	SL 0	1/2,1/2,1/2	Fm$\overline{3}$m (4a,b)		
4 a 23	SL 0	0,0,0	Fm$\overline{3}$m (4a,b)		

Space group No.197 I23

Superlattices

SL 0 I(1,1,1)   density 1 add.gen.
SL 1 P(1/2,1/2,1/2) density 4 add.gen. 1/2,0,0 0,1/2,0

Wyckoff letter   non-characteristic crystallographic orbits

24 f 1	SL 0	0,y,z	Im$\overline{3}$ (24g)	x,x,z	I$\overline{4}$3m (24g)
		1/4,x,1/2-x	I432 (24i)	0,x,x	Im$\overline{3}$m (24h)
	SL 1	1/4,1/4,0	Pm$\overline{3}$m (3c,d)		
12 e 2	SL 0	x,1/2,0	Im$\overline{3}$ (12e)	1/4,1/2,0	Im$\overline{3}$m (12d)
12 d 2	SL 0	x,0,0	Im$\overline{3}$m (12e)		
8 c 3	SL 0	x,x,x	I$\overline{4}$3m (8c)		
	SL 1	1/4,1/4,1/4	Pm$\overline{3}$m (1a,b)		
6 b 222	SL 0	0,1/2,1/2	Im$\overline{3}$m (6b)		
2 a 23	SL 0	0,0,0	Im$\overline{3}$m (2a)		

Space group No.198 P2$_1$3

Superlattices

SL 0 P(1,1,1) density 1 add.gen.
SL 1 I(1,1,1) density 2 add.gen. 1/2,1/2,1/2
SL 2 F(1,1,1) density 4 add.gen. 0,1/2,1/2 1/2,0,1/2

Wyckoff letter   non-characteristic crystallographic orbits

12 b 1	SL 0	1/8,x,1/4-x	P4$_3$32 (12d)	1/8,x,1/4+x	P4$_1$32 (12d)
	SL 1	x,0,1/4	I2$_1$3 (12b)	1/8,0,1/4	I4$_1$32 (12c,d)
		5/8,0,1/4	I4$_1$32 (12c,d)	3/8,0,1/4	I$\overline{4}$3d (12a,b)
		7/8,0,1/4	I$\overline{4}$3d (12a,b)		
4 a 3	SL 0	5/8,5/8,5/8	P4$_3$32 (4a,b)	1/8,1/8,1/8	P4$_3$32 (4a,b)
		7/8,7/8,7/8	P4$_1$32 (4a,b)	3/8,3/8,3/8	P4$_1$32 (4a,b)
	SL 2	0,0,0	Fm$\overline{3}$m (4a,b)	3/4,3/4,3/4	Fm$\overline{3}$m (4a,b)
		1/2,1/2,1/2	Fm$\overline{3}$m (4a,b)	1/4,1/4,1/4	Fm$\overline{3}$m (4a,b)

Space group No.199 $I2_13$

Superlattices

SL	0	I(1,1,1)	density	1	add.gen.	
SL	1	P(1/2,1/2,1/2)	density	4	add.gen.	1/2,0,0 0,1/2,0

Wyckoff letter		non-characteristic crystallographic orbits			
24 c 1	SL 0	1/8,x,1/4-x	$I4_132$ (24g,h)	1/8,x,1/4+x	$I4_132$ (24g,h)
12 b 2	SL 0	7/8,0,1/4	$I\bar{4}3d$ (12a,b)	3/8,0,1/4	$I\bar{4}3d$ (12a,b)
		1/8,0,1/4	$I4_132$ (12c,d)	5/8,0,1/4	$I4_132$ (12c,d)
8 a 3	SL 0	7/8,7/8,7/8	$I4_132$ (8a,b)	1/8,1/8,1/8	$I4_132$ (8a,b)
	SL 1	0,0,0	$Pm\bar{3}m$ (1a,b)	1/4,1/4,1/4	$Pm\bar{3}m$ (1a,b)

Space group No.200 $Pm\bar{3}$

Superlattices

SL	0	P(1,1,1)	density	1	add.gen.	
SL	3	P(1/2,1/2,1/2)	density	8	add.gen.	1/2,0,0 0,1/2,0 0,0,1/2

Wyckoff letter		non-characteristic crystallographic orbits			
24 l 1	SL 0	x,x,z	$Pm\bar{3}m$ (24m)	1/4,x,1/2+x	$Pm\bar{3}n$ (24j)
12 k m	SL 0	1/2,x,x	$Pm\bar{3}m$ (12i,j)		
12 j m	SL 0	0,x,x	$Pm\bar{3}m$ (12i,j)		
8 i 3	SL 0	x,x,x	$Pm\bar{3}m$ (8g)		
	SL 3	1/4,1/4,1/4	$Pm\bar{3}m$ (1a,b)		
6 h mm2	SL 0	x,1/2,1/2	$Pm\bar{3}m$ (6e,f)		
6 g mm2	SL 0	1/4,1/2,0	$Pm\bar{3}n$ (6c,d)		
6 f mm2	SL 0	1/4,0,1/2	$Pm\bar{3}n$ (6c,d)		
6 e mm2	SL 0	x,0,0	$Pm\bar{3}m$ (6e,f)		
3 d mmm	SL 0	1/2,0,0	$Pm\bar{3}m$ (3c,d)		
3 c mmm	SL 0	0,1/2,1/2	$Pm\bar{3}m$ (3c,d)		
1 b m3	SL 0	1/2,1/2,1/2	$Pm\bar{3}m$ (1a,b)		
1 a m3	SL 0	0,0,0	$Pm\bar{3}m$ (1a,b)		

Space group No.201 Pn$\overline{3}$

Superlattices

```
SL  0  P(1,1,1)         density  1  add.gen.
SL  1  I(1,1,1)         density  2  add.gen.  1/2,1/2,1/2
SL  2  F(1,1,1)         density  4  add.gen.  0,1/2,1/2  1/2,0,1/2
SL  3  P(1/2,1/2,1/2)   density  8  add.gen.  1/2,0,0  0,1/2,0  0,0,1/2
```

Wyckoff letter   non-characteristic crystallographic orbits

```
24 h 1     SL 0   x,x,z         Pn3m (24k)      1/2,x,1/2+x  Pn3m (24i,j)
                  1/2,x,-x      Pn3m (24i,j)
           SL 1   1/4,y,z       Im3 (24g)       1/4,x,x      Im3m (24h)
           SL 2   x,0,0         Fm3m (24e)
           SL 3   1/4,0,0       Pm3m (3c,d)
12 g 2     SL 1   x,3/4,1/4     Im3 (12e)       0,3/4,1/4    Im3m (12d)
12 f 2     SL 1   x,1/4,1/4     Im3m (12e)
 8 e 3     SL 0   x,x,x         Pn3m (8e)
 6 d 222   SL 1   1/4,3/4,3/4   Im3m (6b)
 4 c 3     SL 2   1/2,1/2,1/2   Fm3m (4a,b)
 4 b 3     SL 2   0,0,0         Fm3m (4a,b)
 2 a 23    SL 1   1/4,1/4,1/4   Im3m (2a)
```

Space group No.202 Fm$\overline{3}$

Superlattices

```
SL  0  F(1,1,1)         density  1  add.gen.
SL  1  P(1/2,1/2,1/2)   density  2  add.gen.  1/2,0,0
```

Wyckoff letter   non-characteristic crystallographic orbits

```
96 i 1     SL 0   x,x,z         Fm3m (96k)
           SL 1   x,1/4,z       Pm3 (12j,k)     x,1/4,x      Pm3m (12i,j)
48 h m     SL 0   0,y,y         Fm3m (48h,i)    0,y,1/2+y    Fm3m (48h,i)
           SL 1   0,y,1/4       Pm3 (6f,g)      0,1/4,z      Pm3 (6f,g)
                  0,1/8,1/4     Pm3n (6c,d)     0,1/4,1/8    Pm3n (6c,d)
48 g 2     SL 1   x,1/4,1/4     Pm3m (6e,f)
32 f 3     SL 0   x,x,x         Fm3m (32f)
24 e mm2   SL 0   x,0,0         Fm3m (24e)
           SL 1   1/4,0,0       Pm3m (3c,d)
24 d 2/m   SL 1   0,1/4,1/4     Pm3m (3c,d)
 8 c 23    SL 1   1/4,1/4,1/4   Pm3m (1a,b)
 4 b m3    SL 0   1/2,1/2,1/2   Fm3m (4a,b)
 4 a m3    SL 0   0,0,0         Fm3m (4a,b)
```

Space group No.203 Fd$\overline{3}$

Superlattices

SL 0 I(1,1,1)   density 1 add.gen.
SL 2 I(1/2,1/2,1/2) density 4 add.gen. 1/4,1/4,1/4
SL 3 F(1/2,1/2,1/2) density 8 add.gen. 1/2,0,0 0,1/4,1/4 1/4,0,1/4

Wyckoff letter   non-characteristic crystallographic orbits

96 g 1	SL 0	0,x,-x	Fd$\overline{3}$m (96h)		x,x,z	Fd$\overline{3}$m (96g)
		1/4,x,-x	Fd$\overline{3}$c (96g)			
	SL 2	x,1/8,3/8	Im$\overline{3}$ (12e)		1/4,1/8,3/8	Im$\overline{3}$m (12d)
48 f 2	SL 0	x,1/8,1/8	Fd$\overline{3}$m (48f)			
	SL 2	7/8,1/8,1/8	Im$\overline{3}$m (6b)			
32 e 3	SL 0	x,x,x	Fd$\overline{3}$m (32e)			
	SL 3	1/4,1/4,1/4	Fm$\overline{3}$m (4a,b)			
16 d $\overline{3}$	SL 0	1/2,1/2,1/2	Fd$\overline{3}$m (16c,d)			
16 c $\overline{3}$	SL 0	0,0,0	Fd$\overline{3}$m (16c,d)			
8 b 23	SL 0	5/8,5/8,5/8	Fd$\overline{3}$m (8a,b)			
8 a 23	SL 0	1/8,1/8,1/8	Fd$\overline{3}$m (8a,b)			

Space group No.204 Im$\overline{3}$

Superlattices

SL 0 I(1,1,1)   density 1 add.gen.
SL 1 P(1/2,1/2,1/2) density 4 add.gen. 1/2,0,0 0,1/2,0

Wyckoff letter   non-characteristic crystallographic orbits

48 h 1	SL 0	x,x,z	Im$\overline{3}$m (48k)	1/4,x,1/2-x	Im$\overline{3}$m (48i)
	SL 1	x,1/4,1/4	Pm$\overline{3}$m (6e,f)		
24 g m	SL 0	0,y,y	Im$\overline{3}$m (24h)		
	SL 1	0,1/4,1/4	Pm$\overline{3}$m (3c,d)		
16 f 3	SL 0	x,x,x	Im$\overline{3}$m (16f)		
12 e mm2	SL 0	1/4,0,1/2	Im$\overline{3}$m (12d)		
12 d mm2	SL 0	x,0,0	Im$\overline{3}$m (12e)		
8 c $\overline{3}$	SL 1	1/4,1/4,1/4	Pm$\overline{3}$m (1a,b)		
6 b mmm	SL 0	0,1/2,1/2	Im$\overline{3}$m (6b)		
2 a m3	SL 0	0,0,0	Im$\overline{3}$m (2a)		

Space group No.205 Pa$\bar{3}$

Superlattices

SL	1	I(1,1,1)	density	2	add.gen.	1/2,1/2,1/2
SL	2	F(1,1,1)	density	4	add.gen.	0,1/2,1/2 1/2,0,1/2
SL	3	P(1/2,1/2,1/2)	density	8	add.gen.	1/2,0,0 0,1/2,0 0,0,1/2

Wyckoff letter   non-characteristic crystallographic orbits

24 d 1	SL 1	1/4,y,0	Ia$\bar{3}$ (24d)	1/4,1/8,0	Ia$\bar{3}$d (24c)
		1/4,3/8,0	Ia$\bar{3}$d (24d)		
	SL 2	0,0,z	Fm$\bar{3}$m (24e)		
	SL 3	0,1/4,0	Pm$\bar{3}$m (3c,d)	1/4,1/4,0	Pm$\bar{3}$m (3c,d)
8 c 3	SL 3	1/4,1/4,1/4	Pm$\bar{3}$m (1a,b)		
4 b $\bar{3}$	SL 2	1/2,1/2,1/2	Fm$\bar{3}$m (4a,b)		
4 a $\bar{3}$	SL 2	0,0,0	Fm$\bar{3}$m (4a,b)		

Space group No.206 Ia$\bar{3}$

Superlattices

SL	0	I(1,1,1)	density	1	add.gen.	
SL	1	P(1/2,1/2,1/2)	density	4	add.gen.	1/2,0,0 0,1/2,0

Wyckoff letter   non-characteristic crystallographic orbits

48 e 1	SL 0	1/8,x,1/4-x	Ia$\bar{3}$d (48g)		
	SL 1	x,1/4,0	Pm$\bar{3}$ (6f,g)	1/8,1/4,0	Pm$\bar{3}$n (6c,d)
		0,0,z	Pm$\bar{3}$m (6e,f)	1/4,1/4,z	Pm$\bar{3}$m (6e,f)
24 d 2	SL 0	3/8,0,1/4	Ia$\bar{3}$d (24d)	1/8,0,1/4	Ia$\bar{3}$d (24c)
	SL 1	0,0,1/4	Pm$\bar{3}$m (3c,d)	1/4,0,1/4	Pm$\bar{3}$m (3c,d)
16 c 3	SL 0	1/8,1/8,1/8	Ia$\bar{3}$d (16b)		
8 b $\bar{3}$	SL 1	1/4,1/4,1/4	Pm$\bar{3}$m (1a,b)		
8 a $\bar{3}$	SL 1	0,0,0	Pm$\bar{3}$m (1a,b)		

Space group   No.207   P432

Superlattices

```
SL  0   P(1,1,1)       density  1  add.gen.
SL  1   I(1,1,1)       density  2  add.gen.   1/2,1/2,1/2
SL  3   P(1/2,1/2,1/2) density  8  add.gen.   1/2,0,0  0,1/2,0  0,0,1/2
```

Wyckoff letter   non-characteristic crystallographic orbits

```
24 k 1      SL 0   x,x,z        Pm3̄m (24m)      1/2,y,z      Pm3̄m (24k,1)
                   0,y,z        Pm3̄m (24k,1)
            SL 1   1/4,y,1/2-y  I432 (24i)
12 j 2      SL 0   1/2,y,y      Pm3̄m (12i,j)
12 i 2      SL 0   0,y,y        Pm3̄m (12i,j)
12 h 2      SL 0   x,1/2,0      Pm3̄m (12h)
            SL 1   1/4,1/2,0    Im3̄m (12d)
 8 g 3      SL 0   x,x,x        Pm3̄m (8g)
            SL 3   1/4,1/4,1/4  Pm3̄m (1a,b)
 6 f 4      SL 0   x,1/2,1/2    Pm3̄m (6e,f)
 6 e 4      SL 0   x,0,0        Pm3̄m (6e,f)
 3 d 42     SL 0   1/2,0,0      Pm3̄m (3c,d)
 3 c 42     SL 0   0,1/2,1/2    Pm3̄m (3c,d)
 1 b 432    SL 0   1/2,1/2,1/2  Pm3̄m (1a,b)
 1 a 432    SL 0   0,0,0        Pm3̄m (1a,b)
```

Space group   No.208   P4₂32

Superlattices

```
SL  0   P(1,1,1)       density  1  add.gen.
SL  1   I(1,1,1)       density  2  add.gen.   1/2,1/2,1/2
SL  2   F(1,1,1)       density  4  add.gen.   0,1/2,1/2  1/2,0,1/2
SL  3   P(1/2,1/2,1/2) density  8  add.gen.   1/2,0,0  0,1/2,0  0,0,1/2
```

Wyckoff letter   non-characteristic crystallographic orbits

```
24 m 1      SL 0   0,y,z        Pm3̄n (24k)      x,x,z        Pn3̄m (24k)
            SL 1   0,y,y        Im3̄m (24h)
            SL 2   x,1/4,1/4    Fm3̄m (24e)
            SL 3   0,1/4,1/4    Pm3̄m (3c,d)
12 j 2      SL 0   x,1/2,0      Pm3̄n (12g,h)
12 i 2      SL 0   x,0,1/2      Pm3̄n (12g,h)
12 h 2      SL 1   x,0,0        Im3̄m (12e)
 8 g 3      SL 0   x,x,x        Pn3̄m (8e)
 6 f 222    SL 0   1/4,1/2,0    Pm3̄n (6c,d)
 6 e 222    SL 0   1/4,0,1/2    Pm3̄n (6c,d)
 6 d 222    SL 1   0,1/2,1/2    Im3̄m (6b)
 4 c 32     SL 2   3/4,3/4,3/4  Fm3̄m (4a,b)
 4 b 32     SL 2   1/4,1/4,1/4  Fm3̄m (4a,b)
 2 a 23     SL 1   0,0,0        Im3̄m (2a)
```

Space group   No.209   F432

Superlattices

```
SL  0  F(1,1,1)          density  1  add.gen.
SL  1  P(1/2,1/2,1/2)    density  2  add.gen.   1/2,0,0
SL  2  I(1/2,1/2,1/2)    density  4  add.gen.   1/4,1/4,1/4
```

Wyckoff letter   non-characteristic crystallographic orbits

```
96 j 1    SL 0   x,x,z         Fm3m (96k)      0,y,z      Fm3m (96j)
                 1/4,y,z       Fm3c (96i)
          SL 1   x,1/4,0       Pm3m (12h)      1/4,y,y    Pm3m (12i,j)
          SL 2   1/4,1/8,0     Im3m (12d)
48 i 2    SL 1   x,1/4,1/4     Pm3m (6e,f)
48 h 2    SL 0   1/2,y,y       Fm3m (48h,i)
48 g 2    SL 0   0,y,y         Fm3m (48h,i)
32 f 3    SL 0   x,x,x         Fm3m (32f)
24 e 4    SL 0   x,0,0         Fm3m (24e)
          SL 1   1/4,0,0       Pm3m (3c,d)
24 d 222  SL 1   0,1/4,1/4     Pm3m (3c,d)
 8 c 23   SL 1   1/4,1/4,1/4   Pm3m (1a,b)
 4 b 432  SL 0   1/2,1/2,1/2   Fm3m (4a,b)
 4 a 432  SL 0   0,0,0         Fm3m (4a,b)
```

Space group   No.210   F4₁32

Superlattices

```
SL  0  F(1,1,1)          density  1  add.gen.
SL  1  P(1/2,1/2,1/2)    density  2  add.gen.   1/2,0,0
SL  2  I(1/2,1/2,1/2)    density  4  add.gen.   1/4,1/4,1/4
SL  3  F(1/2,1/2,1/2)    density  8  add.gen.   1/2,0,0  0,1/4,1/4  1/4,0,1/4
```

Wyckoff letter   non-characteristic crystallographic orbits

```
96 h 1    SL 0   x,x,z         Fd3m (96g)
          SL 1   1/8,x,1/4+x   P4₂32 (12k,l)    x,0,1/4    Pm3n (12g,h)
                 x,1/4,0       Pm3n (12g,h)
48 g 2    SL 1   1/8,3/4,1/2   Pm3n (6c,d)      1/8,0,1/4  Pm3n (6c,d)
48 f 2    SL 0   x,0,0         Fd3m (48f)
          SL 2   1/4,0,0       Im3m (6b)
32 e 3    SL 0   x,x,x         Fd3m (32e)
          SL 3   3/8,3/8,3/8   Fm3m (4a,b)
16 d 32   SL 0   5/8,5/8,5/8   Fd3m (16c,d)
16 c 32   SL 0   1/8,1/8,1/8   Fd3m (16c,d)
 8 b 23   SL 0   1/2,1/2,1/2   Fd3m (8a,b)
 8 a 23   SL 0   0,0,0         Fd3m (8a,b)
```

Space group   No.211   I432

Superlattices

```
SL  0  I(1,1,1)         density  1  add.gen.
SL  1  P(1/2,1/2,1/2)   density  4  add.gen.   1/2,0,0   0,1/2,0
```

Wyckoff letter   non-characteristic crystallographic orbits

```
48 j 1      SL 0   x,x,z       Im3m (48k)      0,y,z      Im3m (48j)
            SL 1   x,1/4,1/4   Pm3m (6e,f)
24 h 2      SL 0   0,y,y       Im3m (24h)
            SL 1   0,1/4,1/4   Pm3m (3c,d)
24 g 2      SL 0   x,1/2,0     Im3m (24g)
16 f 3      SL 0   x,x,x       Im3m (16f)
12 e 4      SL 0   x,0,0       Im3m (12e)
12 d 222    SL 0   1/4,1/2,0   Im3m (12d)
 8 c 32     SL 1   1/4,1/4,1/4 Pm3m (1a,b)
 6 b 42     SL 0   0,1/2,1/2   Im3m (6b)
 2 a 432    SL 0   0,0,0       Im3m (2a)
```

Space group   No.212   P4₃32

Superlattices

```
SL  1  I(1,1,1)         density  2  add.gen.   1/2,1/2,1/2
SL  2  F(1,1,1)         density  4  add.gen.   0,1/2,1/2   1/2,0,1/2
```

Wyckoff letter   non-characteristic crystallographic orbits

```
24 e 1      SL 1   x,0,1/4     I4₁32 (24f)      1/8,x,1/4+x   I4₁32 (24g,h)
                   3/8,0,1/4   Ia3d (24d)
12 d 2      SL 1   1/8,1/2,3/4 I4₁32 (12c,d)    1/8,0,1/4     I4₁32 (12c,d)
 8 c 3      SL 1   7/8,7/8,7/8 I4₁32 (8a,b)
            SL 2   0,0,0       Fd3m (8a,b)      1/2,1/2,1/2   Fd3m (8a,b)
```

Space group No.213 $P4_132$

Superlattices

SL 1	I(1,1,1)	density 2	add.gen.	1/2,1/2,1/2
SL 2	F(1,1,1)	density 4	add.gen.	0,1/2,1/2 1/2,0,1/2

Wyckoff letter non-characteristic crystallographic orbits

24 e 1	SL 1	x,0,1/4	$I4_132$ (24f)	1/8,x,1/4-x	$I4_132$ (24g,h)
		3/8,0,1/4	$Ia\bar{3}d$ (24d)		
12 d 2	SL 1	1/8,1/2,3/4	$I4_132$ (12c,d)	1/8,0,1/4	$I4_132$ (12c,d)
8 c 3	SL 1	1/8,1/8,1/8	$I4_132$ (8a,b)		
	SL 2	0,0,0	$Fd\bar{3}m$ (8a,b)	1/2,1/2,1/2	$Fd\bar{3}m$ (8a,b)

Space group No.214 $I4_132$

Superlattices

SL 0	I(1,1,1)	density 1	add.gen.	
SL 2	I(1/2,1/2,1/2)	density 8	add.gen.	1/2,0,0 0,1/2,0 1/4,1/4,1/4

Wyckoff letter non-characteristic crystallographic orbits

24 f 2	SL 0	3/8,0,1/4	$Ia\bar{3}d$ (24d)
16 e 3	SL 2	0,0,0	$Im\bar{3}m$ (2a)

Space group No.215 P4̄3m

Superlattices

```
SL  0  P(1,1,1)        density  1  add.gen.
SL  1  I(1,1,1)        density  2  add.gen.  1/2,1/2,1/2
SL  2  F(1,1,1)        density  4  add.gen.  0,1/2,1/2  1/2,0,1/2
```

Wyckoff letter   non-characteristic crystallographic orbits

```
 24 j 1      SL 0   0,y,z           Pm3̄m (24k,l)    1/2,y,z         Pm3̄m (24k,l)
                    1/4,x,1/2+x     Pn3̄m (24i,j)    1/4,x,1/2-x     Pn3̄m (24i,j)
 12 i m      SL 0   x,x,1/2         Pm3̄m (12i,j)    x,x,0           Pm3̄m (12i,j)
 12 h 2      SL 0   x,1/2,0         Pm3̄m (12h)
             SL 1   1/4,1/2,0       Im3̄m (12d)
  6 g mm2    SL 0   x,1/2,1/2       Pm3̄m (6e,f)
  6 f mm2    SL 0   x,0,0           Pm3̄m (6e,f)
  4 e 3m     SL 2   3/4,3/4,3/4     Fm3̄m (4a,b)     1/4,1/4,1/4     Fm3̄m (4a,b)
  3 d 4̄2m    SL 0   1/2,0,0         Pm3̄m (3c,d)
  3 c 4̄2m    SL 0   0,1/2,1/2       Pm3̄m (3c,d)
  1 b 4̄3m    SL 0   1/2,1/2,1/2     Pm3̄m (1a,b)
  1 a 4̄3m    SL 0   0,0,0           Pm3̄m (1a,b)
```

Space group No.216 F4̄3m

Superlattices

```
SL  0  F(1,1,1)          density  1  add.gen.
SL  1  P(1/2,1/2,1/2)    density  2  add.gen.  1/2,0,0
SL  2  I(1/2,1/2,1/2)    density  4  add.gen.  1/4,1/4,1/4
```

Wyckoff letter   non-characteristic crystallographic orbits

```
 96 i 1      SL 0   0,y,z           Fm3̄m (96j)      1/4,y,z         Fm3̄m (96j)
                    1/8,x,1/4-x     Fd3̄m (96h)      3/8,x,3/4-x     Fd3̄m (96h)
             SL 1   x,1/4,0         Pm3̄m (12h)
             SL 2   1/8,1/4,0       Im3̄m (12d)
 48 h m      SL 0   x,x,1/2         Fm3̄m (48h,i)    x,x,1/4         Fm3̄m (48h,i)
                    x,x,0           Fm3̄m (48h,i)    x,x,3/4         Fm3̄m (48h,i)
 24 g mm2    SL 0   x,1/4,1/4       Fm3̄m (24e)
             SL 1   0,1/4,1/4       Pm3̄m (3c,d)
 24 f mm2    SL 0   x,0,0           Fm3̄m (24e)
             SL 1   1/4,0,0         Pm3̄m (3c,d)
 16 e 3m     SL 0   5/8,5/8,5/8     Fd3̄m (16c,d)    7/8,7/8,7/8     Fd3̄m (16c,d)
                    3/8,3/8,3/8     Fd3̄m (16c,d)    1/8,1/8,1/8     Fd3̄m (16c,d)
  4 d 4̄3m    SL 0   3/4,3/4,3/4     Fm3̄m (4a,b)
  4 c 4̄3m    SL 0   1/4,1/4,1/4     Fm3̄m (4a,b)
  4 b 4̄3m    SL 0   1/2,1/2,1/2     Fm3̄m (4a,b)
  4 a 4̄3m    SL 0   0,0,0           Fm3̄m (4a,b)
```

Space group   No.217   I$\bar{4}$3m

Superlattices

```
SL  0  I(1,1,1)          density  1  add.gen.
SL  1  P(1/2,1/2,1/2)  density  4  add.gen.  1/2,0,0   0,1/2,0
```

Wyckoff letter   non-characteristic crystallographic orbits

```
48 h 1      SL 0   0,y,z         Im3m (48j)      1/4,x,1/2-x  Im3m (48i)
24 g m      SL 0   x,x,0         Im3m (24h)
            SL 1   1/4,1/4,0     Pm3m (3c,d)
24 f 2      SL 0   x,1/2,0       Im3m (24g)
12 e mm2    SL 0   x,0,0         Im3m (12e)
12 d 4      SL 0   1/4,1/2,0     Im3m (12d)
 8 c 3m     SL 1   1/4,1/4,1/4   Pm3m (1a,b)
 6 b 42m    SL 0   0,1/2,1/2     Im3m (6b)
 2 a 43m    SL 0   0,0,0         Im3m (2a)
```

Space group   No.218   P$\bar{4}$3n

Superlattices

```
SL  0  P(1,1,1)          density  1  add.gen.
SL  1  I(1,1,1)          density  2  add.gen.  1/2,1/2,1/2
SL  3  P(1/2,1/2,1/2)  density  8  add.gen.  1/2,0,0  0,1/2,0  0,0,1/2
```

Wyckoff letter   non-characteristic crystallographic orbits

```
24 i 1      SL 0   0,y,z         Pm3n (24k)      1/4,x,1/2+x  Pm3n (24j)
            SL 1   x,x,z         I43m (24g)      0,x,x        Im3m (24h)
            SL 3   1/4,1/4,0     Pm3m (3c,d)
12 h 2      SL 0   x,0,1/2       Pm3n (12g,h)
12 g 2      SL 0   x,1/2,0       Pm3n (12g,h)
12 f 2      SL 1   x,0,0         Im3m (12e)
 8 e 3      SL 1   x,x,x         I43m (8c)
            SL 3   1/4,1/4,1/4   Pm3m (1a,b)
 6 d 4      SL 0   1/4,0,1/2     Pm3n (6c,d)
 6 c 4      SL 0   1/4,1/2,0     Pm3n (6c,d)
 6 b 222    SL 1   0,1/2,1/2     Im3m (6b)
 2 a 23     SL 1   0,0,0         Im3m (2a)
```

Space group   No.219   F$\bar{4}$3c

Superlattices

```
SL  0  F(1,1,1)        density  1  add.gen.
SL  1  P(1/2,1/2,1/2)  density  2  add.gen.  1/2,0,0
SL  2  I(1/2,1/2,1/2)  density  4  add.gen.  1/4,1/4,1/4
SL  3  F(1/2,1/2,1/2)  density  8  add.gen.  1/2,0,0  0,1/4,1/4  1/4,0,1/4
```

Wyckoff letter   non-characteristic crystallographic orbits

```
96 h 1     SL 0   0,y,z          Fm3c (96i)      1/4,y,z        Fm3c (96i)
                  1/8,x,1/4-x    Fd3c (96g)      3/8,x,3/4-x    Fd3c (96g)
           SL 1   x,x,z          P43m (12i)      x,x,0          Pm3m (12i,j)
                  x,x,1/4        Pm3m (12i,j)    x,1/4,0        Pm3m (12h)
           SL 2   1/8,0,1/4      Im3m (12d)
48 g 2     SL 1   x,1/4,1/4      Pm3m (6e,f)
48 f 2     SL 1   x,0,0          Pm3m (6e,f)
32 e 3     SL 1   x,x,x          P43m (4e)
           SL 3   3/8,3/8,3/8    Fm3m (4a,b)     1/8,1/8,1/8    Fm3m (4a,b)
24 d 4     SL 1   1/4,0,0        Pm3m (3c,d)
24 c 4     SL 1   0,1/4,1/4      Pm3m (3c,d)
 8 b 23    SL 1   1/4,1/4,1/4    Pm3m (1a,b)
 8 a 23    SL 1   0,0,0          Pm3m (1a,b)
```

Space group   No.220   I$\bar{4}$3d

Superlattices

```
SL  0  I(1,1,1)        density  1  add.gen.
SL  1  P(1/2,1/2,1/2)  density  4  add.gen.  1/2,0,0  0,1/2,0
SL  2  I(1/2,1/2,1/2)  density  8  add.gen.  1/2,0,0  0,1/2,0  1/4,1/4,1/4
```

Wyckoff letter   non-characteristic crystallographic orbits

```
48 e 1     SL 0   1/8,x,1/4-x    Ia3d (48g)
           SL 1   1/8,1/4,0      Pm3n (6c,d)
24 d 2     SL 0   1/8,0,1/4      Ia3d (24c)
16 c 3     SL 0   1/8,1/8,1/8    Ia3d (16b)
           SL 2   0,0,0          Im3m (2a)
```

Space group   No.221   Pm$\overline{3}$m

Superlattices

```
SL  1  I(1,1,1)        density  2  add.gen.  1/2,1/2,1/2
SL  3  P(1/2,1/2,1/2)  density  8  add.gen.  1/2,0,0  0,1/2,0  0,0,1/2
```

Wyckoff letter   non-characteristic crystallographic orbits

```
48 n 1     SL 1  1/4,x,1/2-x  Im3m (48i)
12 h mm2   SL 1  1/4,1/2,0    Im3m (12d)
 8 g 3m    SL 3  1/4,1/4,1/4  Pm3m (1a,b)
```

Space group   No.222   Pn$\overline{3}$n

Superlattices

```
SL  1  I(1,1,1)        density  2  add.gen.  1/2,1/2,1/2
SL  3  P(1/2,1/2,1/2)  density  8  add.gen.  1/2,0,0  0,1/2,0  0,0,1/2
```

Wyckoff letter   non-characteristic crystallographic orbits

```
48 i 1     SL 1  1/4,y,z        Im3m (48j)     x,x,z      Im3m (48k)
                 x,-x,0         Im3m (48i)
           SL 3  x,0,0          Pm3m (6e,f)
24 h 2     SL 1  1/4,y,y        Im3m (24h)
           SL 3  1/4,0,0        Pm3m (3c,d)
24 g 2     SL 1  x,3/4,1/4      Im3m (24g)
16 f 3     SL 1  x,x,x          Im3m (16f)
12 e 4     SL 1  x,1/4,1/4      Im3m (12e)
12 d 4̄     SL 1  0,3/4,1/4      Im3m (12d)
 8 c 3̄     SL 3  0,0,0          Pm3m (1a,b)
 6 b 42    SL 1  3/4,1/4,1/4    Im3m (6b)
 2 a 432   SL 1  1/4,1/4,1/4    Im3m (2a)
```

Space group   No.223   Pm$\bar{3}$n

Superlattices

SL 1 I(1,1,1)   density 2 add.gen. 1/2,1/2,1/2
SL 3 P(1/2,1/2,1/2) density 8 add.gen. 1/2,0,0 0,1/2,0 0,0,1/2

Wyckoff letter   non-characteristic crystallographic orbits

```
48  l  1      SL 1   x,x,z          Im3m (48k)
              SL 3   x,1/4,1/4      Pm3m (6e,f)
24  k  m      SL 1   0,y,y          Im3m (24h)
              SL 3   0,1/4,1/4      Pm3m (3c,d)
16  i  3      SL 1   x,x,x          Im3m (16f)
12  f  mm2    SL 1   x,0,0          Im3m (12e)
 8  e  32     SL 3   1/4,1/4,1/4    Pm3m (1a,b)
 6  b  mmm    SL 1   0,1/2,1/2      Im3m (6b)
 2  a  m3     SL 1   0,0,0          Im3m (2a)
```

Space group   No.224   Pn$\bar{3}$m

Superlattices

SL 1 I(1,1,1)   density 2 add.gen. 1/2,1/2,1/2
SL 2 F(1,1,1)   density 4 add.gen. 0,1/2,1/2 1/2,0,1/2
SL 3 P(1/2,1/2,1/2) density 8 add.gen. 1/2,0,0 0,1/2,0 0,0,1/2

Wyckoff letter   non-characteristic crystallographic orbits

```
48  l  1      SL 1   1/4,y,z        Im3m (48j)
24  k  m      SL 1   x,x,1/4        Im3m (24h)
              SL 2   0,0,z          Fm3m (24e)
              SL 3   0,0,1/4        Pm3m (3c,d)
24  h  2      SL 1   x,1/4,3/4      Im3m (24g)
12  g  mm2    SL 1   x,1/4,1/4      Im3m (12e)
12  f  222    SL 1   1/2,1/4,3/4    Im3m (12d)
 6  d  42m    SL 1   1/4,3/4,3/4    Im3m (6b)
 4  c  3m     SL 2   1/2,1/2,1/2    Fm3m (4a,b)
 4  b  3m     SL 2   0,0,0          Fm3m (4a,b)
 2  a  43m    SL 1   1/4,1/4,1/4    Im3m (2a)
```

Space group No.225 Fm$\bar{3}$m

Superlattices

```
SL  1  P(1/2,1/2,1/2)  density  2  add.gen.  1/2,0,0
SL  2  I(1/2,1/2,1/2)  density  4  add.gen.  1/4,1/4,1/4
```

Wyckoff letter   non-characteristic crystallographic orbits

```
192 l 1     SL 1  1/4,y,z        Pm3m (24k,l)
 96 k m     SL 1  x,x,1/4        Pm3m (12i,j)
 96 j m     SL 1  0,1/4,z        Pm3m (12h)
            SL 2  0,1/4,1/8      Im3m (12d)
 48 g mm2   SL 1  x,1/4,1/4      Pm3m (6e,f)
 24 e 4mm   SL 1  1/4,0,0        Pm3m (3c,d)
 24 d mmm   SL 1  0,1/4,1/4      Pm3m (3c,d)
  8 c 43m   SL 1  1/4,1/4,1/4    Pm3m (1a,b)
```

Space group No.226 Fm$\bar{3}$c

Superlattices

```
SL  1  P(1/2,1/2,1/2)  density   2  add.gen.  1/2,0,0
SL  2  I(1/2,1/2,1/2)  density   4  add.gen.  1/4,1/4,1/4
SL  4  P(1/4,1/4,1/4)  density  16  add.gen.  1/4,0,0  0,1/4,0  0,0,1/4
```

Wyckoff letter   non-characteristic crystallographic orbits

```
192 j 1     SL 1  1/4,y,z        Pm3m (24k,l)    x,x,z    Pm3m (24m)
 96 i m     SL 1  0,y,1/4        Pm3m (12h)      0,y,y    Pm3m (12i,j)
            SL 2  0,1/8,1/4      Im3m (12d)
 96 h 2     SL 1  1/4,y,y        Pm3m (12i,j)
 64 g 3     SL 1  x,x,x          Pm3m (8g)
            SL 4  1/8,1/8,1/8    Pm3m (1a,b)
 48 f 4     SL 1  x,1/4,1/4      Pm3m (6e,f)
 48 e mm2   SL 1  x,0,0          Pm3m (6e,f)
 24 d 4/m   SL 1  0,1/4,1/4      Pm3m (3c,d)
 24 c 4m2   SL 1  1/4,0,0        Pm3m (3c,d)
  8 b m3    SL 1  0,0,0          Pm3m (1a,b)
  8 a 432   SL 1  1/4,1/4,1/4    Pm3m (1a,b)
```

Space group No.227 Fd$\overline{3}$m

Superlattices

```
SL  1  P(1/2,1/2,1/2)  density  2  add.gen.  1/2,0,0
SL  2  I(1/2,1/2,1/2)  density  4  add.gen.  1/4,1/4,1/4
SL  3  F(1/2,1/2,1/2)  density  8  add.gen.  1/2,0,0  0,1/4,1/4  1/4,0,1/4
```

Wyckoff letter   non-characteristic crystallographic orbits

```
192 i 1      SL 1   1/4,x,1/2-x   Pn3̄m (24i,j)
             SL 2   x,1/8,3/8     Im3̄m (24g)
 96 h 2      SL 2   0,5/8,3/8     Im3̄m (12d)
 48 f mm2    SL 2   3/8,1/8,1/8   Im3̄m (6b)
 32 e 3m     SL 3   1/4,1/4,1/4   Fm3̄m (4a,b)
```

Space group No.228 Fd$\overline{3}$c

Superlattices

```
SL  1  P(1/2,1/2,1/2)  density   2  add.gen.  1/2,0,0
SL  2  I(1/2,1/2,1/2)  density   4  add.gen.  1/4,1/4,1/4
SL  3  F(1/2,1/2,1/2)  density   8  add.gen.  1/2,0,0  0,1/4,1/4  1/4,0,1/4
SL  4  P(1/4,1/4,1/4)  density  16  add.gen.  1/4,0,0  0,1/4,0  0,0,1/4
```

Wyckoff letter   non-characteristic crystallographic orbits

```
192 h 1      SL 1   x,x,z         Pn3̄m (24k)      1/4,x,1/4+x  Pn3̄m (24i,j)
             SL 2   x,1/8,3/8     Im3̄m (24g)      1/8,x,x      Im3̄m (24h)
             SL 3   x,0,0         Fm3̄m (24e)
             SL 4   1/8,0,0       Pn3̄m (3c,d)
 96 g 2      SL 2   1/4,7/8,1/8   Im3̄m (12d)
 96 f 2      SL 2   x,1/8,1/8     Im3̄m (12e)
 64 e 3      SL 1   x,x,x         Pn3̄m (8e)
 48 d 4̄      SL 2   7/8,1/8,1/8   Im3̄m (6b)
 32 c 3̄      SL 3   0,0,0         Fm3̄m (4a,b)
 32 b 32     SL 3   1/4,1/4,1/4   Fm3̄m (4a,b)
 16 a 23     SL 2   1/8,1/8,1/8   Im3̄m (2a)
```

Space group   No.229   Im$\bar{3}$m

Superlattices

SL 1 P(1/2,1/2,1/2) density 4 add.gen.   1/2,0,0   0,1/2,0

Wyckoff letter   non-characteristic crystallographic orbits

```
48 k m      SL 1   1/4,1/4,z     Pm3m (6e,f)
24 h mm2    SL 1   0,1/4,1/4     Pm3m (3c,d)
 8 c 3m     SL 1   1/4,1/4,1/4   Pm3m (1a,b)
```

Space group   No.230   Ia$\bar{3}$d

Superlattices

SL 1 P(1/2,1/2,1/2) density 4 add.gen.   1/2,0,0   0,1/2,0
SL 2 I(1/2,1/2,1/2) density 8 add.gen.   1/2,0,0   0,1/2,0   1/4,1/4,1/4

Wyckoff letter   non-characteristic crystallographic orbits

```
96 h 1      SL 1   x,1/4,0       Pm3n (12g,h)
            SL 2   x,0,0         Im3m (12e)
48 g 2      SL 1   1/8,1/4,0     Pm3n (6c,d)
48 f 2      SL 2   1/4,0,1/4     Im3m (6b)
16 a 3      SL 2   0,0,0         Im3m (2a)
```

4. Comments to the Main Table

In the Main Table those non-characteristic crystallographic orbits are listed whose eigensymmetry
space groups belong to the same crystal family as
the generating space group, thus no accidental higher lattice symmetry is accounted for. For each
space-group type the head line shows the conventional number of the space-group type according to
IT 1983 together with the corresponding
Hermann-Mauguin symbol. The setting of the
space-group types is taken from IT 1983. For the
monoclinic crystal system only the unique axis b was
used. For the rhombohedral space groups the hexagonal setting is taken throughout. In all cases where
two origin choices exist, the one with the centre of
symmetry at the origin was chosen.

In the list of superlattices, for each superlattice
the number corresponding to the numbering in Table 2
and the Bravais letter indicating the lattice type
together with the transformation (f_{11}, f_{22}, f_{33}) of
the basis vectors are given. The three coeffi-
cients, f_{11}, f_{22}, f_{33}, describe the transformation of
the basis vectors, $\vec{a}, \vec{b}, \vec{c}$, of the original lattice
into the basis vectors, $\vec{a}', \vec{b}', \vec{c}'$, of the superlattice. The coefficients f_{ii} are rational numbers

$$\begin{vmatrix} \vec{a}' \\ \vec{b}' \\ \vec{c}' \end{vmatrix} = \begin{vmatrix} f_{11} & 0 & 0 \\ 0 & f_{22} & 0 \\ 0 & 0 & f_{33} \end{vmatrix} \begin{vmatrix} \vec{a} \\ \vec{b} \\ \vec{c} \end{vmatrix} .$$

In the monoclinic C space groups the transformation
$(f_{11}, f_{22}, f_{33}^{*})$ stands for the matrix

$$\begin{vmatrix} f_{11} & 0 & 0 \\ 0 & f_{22} & 0 \\ 1/2f_{33} & 0 & 1/2f_{33} \end{vmatrix} .$$

An example is space group No. 15, C2/c, superlattice
13, C(1,1,1/2*). The transformation matrix writes:

$$\begin{vmatrix} 1 & 0 & 0 \\ 0 & 1 & 0 \\ 1/4 & 0 & 1/4 \end{vmatrix} .$$

In the trigonal space groups the transformation (f_{11}, f_{22}, f_{33}) stands for the matrix

$$\begin{pmatrix} 2/3f_{11} & 1/3f_{11} & 0 \\ -1/3f_{11} & 1/3f_{11} & 0 \\ 0 & 0 & f_{33} \end{pmatrix}$$

An example is space group No. 148, $R\bar{3}$, superlattice 2, $P(1*,1*,1/3)$. The transformation matrix is then

$$\begin{pmatrix} 2/3 & 1/3 & 0 \\ -1/3 & 1/3 & 0 \\ 0 & 0 & 1/3 \end{pmatrix}$$

The Bravais letters RO and RR mean obverse and reverse setting of the rhombohedral lattice with respect to the original hexagonal basis vectors.

If possible the Bravais lattice type is referred to the original basis vectors although non-standard lattice types may result. For example, in the monoclinic space groups the non-standard Bravais lattice types A, B, I, F, occur. Similarly in the tetragonal space groups the Bravais types C and F are non-standard.

The index of the original lattice in the superlattice is called the density and is given together with a set of additional lattice translations which generate the superlattice from the original lattice.

Next follows the list of non-characteristic crystallographic orbits. Each Wyckoff position of the generating space group G which contains non-characteristic crystallographic orbits is given with the multiplicity and the Wyckoff letter, together with the international symbol of the site-symmetry group. The non-characteristic crystallographic orbits are arranged with increasing number of the superlattice. For each eigensymmetry space group E_0 the complete manifold M of all representing points is given whose crystallographic orbits have eigensymmetry space group $E \geq E_0$. Those crystallographic orbits with eigensymmetry space group $E > E_0$ are listed under subsequent entries. They have to be taken away from M in order to obtain the crystallographic orbits with eigensymmetry E_0.

Example. Space group Pmm2, Wyckoff position 4 i 1.
All orbits x,y,z belong to E_0=Pmmm and SL 0 with the
exception of the orbits 1/4,y,z and x,1/4,z. All
orbits 1/4,y,z belong to SL 1 with the exception of
the orbits 1/4,1/4,z, and all orbits x,1/4,z belong
to SL 2 with the exception of the orbits 1/4,1/4,z.
The orbits 1/4,1/4,z belong to SL 9.

For every $X_0 \in M$, the crystallographic orbit $O_G(X_0)$ is
obtained by applying all symmetry operations of G
onto X_0. The dimension of the manifold M can be
taken from the number of different variables, e.g.
1/4,1/2,0, x,1/2-x,1/4, and x,x,z have 0,1, and 2
variables respectively. The number of variable coordinates of the Wyckoff position can be taken from
the symbol of the site-symmetry group as shown
below:

site-symmetry group	number of variables
$m\bar{3}m$, $\bar{4}3m$, 432, $m\bar{3}$, 23, 6/mmm, $\bar{6}m2$, 622, 6/m, $\bar{6}$, $\bar{3}m$, 32, $\bar{3}$, 4/mmm, $\bar{4}2m$, 422, 4/m, $\bar{4}$, mmm, 222, 2/m, $\bar{1}$	0
6mm, 6, 3m, 3, 4mm, 4, mm2, 2	1
m	2
1	3

If the number of variables of M is less than that of
the Wyckoff position then M corresponds to a limiting form as defined by Koch (1974). Not considered
in these tables are eigensymmetries caused by
special lattice parameters. For instance, in the
orthorhombic space group Pnma the position
8 d 1/4,0,0 has eigensymmetry $Pm\bar{3}m$ if all three
lattice parameters are equal in length.

For each crystallographic orbit the eigensymmetry
space group E and the complete symbol for the Wyckoff set in E are given. As the Wyckoff position in
E is not uniquely determined, all Wyckoff positions
within the Wyckoff set are stated.

All non-characteristic crystallographic orbits with
one of the following eigensymmetries, P$\bar{1}$(1a-h),
P2/m(1a-h), C2/m(2a-d), A2/m(2a-d), I2/m(2a-d),
Pmmm(1a-h), Cmmm(2a-d), Ammm(2a-d), Bmmm(2a-d),
Immm(2a-d), Fmmm(4a,b), P4/mmm(1a-d), I4/mmm(2a,b),
R$\bar{3}$m(3a,b), P6/mmm(1a,b), Pm$\bar{3}$m(1a,b), Im$\bar{3}$m(2a),
Fm$\bar{3}$m(4a,b) are point lattices (cf. section 2.4).

In 3-dimensional space only space groups of 166+11
types occur as eigensymmetry space groups of crystallographic orbits, compared to a total of 219+11
space-group types. The corresponding numbers in
2-dimensional space are 13 types of eigensymmetry
plane groups against a total of 17 plane-group types
(the list of these 13 types is given in chapter 7).

There are 1694+37 entries in IT 1983 for Wyckoff
positions which are collected into 1128+29 Wyckoff
sets. Of these, 64+3 types of Wyckoff sets contain
only characteristic, 338+22 contain both characteristic and non-characteristic and 726+4 contain sole-
ly non-characteristic crystallographic orbits. The
first two kinds, 402+25 types of Wyckoff sets determine the 402 lattice complexes (in the definition of
"lattice complex" equivalence under A, not under A$^+$
is used).

Wyckoff positions with only characteristic orbits
are not mentioned in the Main Table. Therefore,
e.g. in space group P4/mmm, the Wyckoff letters l-o
and a-d are missing. There is no 3-dimensional
space group with characteristic crystallographic
orbits only.

5. Determination of non-characteristic crystallographic orbits

5.1. The algebraic approach

For the determination of all extraorbits the algebraic approach proposed by Wondratschek (1976) may
be used. Let G be the generating space group with
lattice L and basis vectors $\vec{a}_1, \ldots, \vec{a}_n$. Let $O_G(X_0)$
be an extraorbit, with eigensymmetry space group E,
lattice L', and basis vectors $\vec{a}_1', \ldots, \vec{a}_n'$. For every
extraorbit the lattice L' is a superlattice of L. We
represent the basis vectors $\vec{a}_i'$ by their components
relative to the basis vectors, $\vec{a}_1, \ldots, \vec{a}_n$.

$$\begin{pmatrix} \vec{a}_1' \\ \cdot \\ \cdot \\ \cdot \\ \vec{a}_n' \end{pmatrix} = \begin{pmatrix} U_{11} & \cdots & U_{1n} \\ \cdot & & \cdot \\ \cdot & & \cdot \\ \cdot & & \cdot \\ U_{n1} & \cdots & U_{nn} \end{pmatrix} \begin{pmatrix} \vec{a}_1 \\ \cdot \\ \cdot \\ \cdot \\ \vec{a}_n \end{pmatrix}$$

The coefficients U_{ij} are fractions and the ratio of
the volumes of the corresponding primitive cells

$$\frac{V}{V'} = \frac{1}{Det(U)} = d \tag{1}$$

is integral and is called the density of the lattice
L' in the lattice L or the index of L in L'. The
density d is equal to the number of additional
translation vectors $\vec{v}_i$ in the primitive cell of L.
This can be seen from the coset decomposition of the
lattice L'

$$L' = L \cup L\vec{v}_2 \cup \ldots \cup L\vec{v}_d.$$

For every $\vec{v}_i \in L'$ and the above-mentioned point X_0
there exists a symmetry operation $g_{0i} \in G$ so that

$$\vec{v}_i = g_{0i}X_0 - X_0 = X_i - X_0 \tag{2}$$

holds. The superlattice L' requires that a similar
equation holds for all points $X_j \in O_G(X_0)$,

$$\vec{v}_i = X_k - X_j = g_{jk}X_j - X_j = g_{jk}g_{0j}X_0 - g_{0j}X_0.$$

It follows that

$$\vec{v}_i' = g_{0j}^{-1}(X_k - X_j) = g_{0j}^{-1}g_{jk}g_{0j}X_0 - X_0 \tag{3}$$

is a translation of L' for all $g_{0j} \in G$. This means the superlattice L' is invariant under all motions of the generating space group G. A superlattice which satisfies condition (3) is called a permitted super-lattice.

Although equation (2) has been stated correctly by Wondratschek (1976), an essential step has been omitted in the derived procedure. On p. 46 of that paper a step 2a has to be inserted in order to check if for all further points of the crystallographic orbit, the above mentioned conditions are fulfilled.

If a superlattice is possible for a crystal family, it may not be permitted for all of its space groups. In 3-dimensional space there exists only one example where condition (3) is important: In space group $P3_1 12$ ($P3_2 12$) the rhombohedral superlattice R is not invariant under the motions of that space group.

For a permitted superlattice L' with additional translation vectors $\vec{v}_i$, i=1,...,d, appropriate symmetry operations $g_{0i} \in G$ have to be determined for all $\vec{v}_i$ and then the system of d equations (2) has to be solved for X_0.

For computational economy first the manifold M_{ij} of points X_k is determined for each translation vector $\vec{v}_i$ and each symmetry operation $g_j \in G$ satisfying equation (2),

$$M_{ij} = \{X_k \mid \vec{v}_i = g_j X_k - X_k\} \qquad (4)$$

If this equation has at least one solution then the dimension of M_{ij} depends only on g_j. In 3-dimensional space the following cases occur:

symmetry operation g_j	dimension of M_{ij}
inversion or roto-inversion	0
rotation or screw rotation	1
reflection or glide reflection	2

It is sufficient to consider solutions within an asymmetric unit F of the Euclidean normalizer $N_E(G)$, cf. section 5.2, of G. Therefore, only those symmetry operations $g_j \in G$ have to be considered for which

$$M_{ij} \cap F \neq \emptyset$$

Finally the appropriate intersections of the M_{ij} give all solutions M for the superlattice L',

$$M = \bigcap_{i=1}^{d} M_{ij}. \qquad (5)$$

These computations require the list of possible superlattices, Matsumoto and Wondratschek (1979). A shortened list which contains only the actually occuring superlattices is shown in Table 2.

In monoclinic space groups with a P lattice all possible superlattices SL 1 to SL 30 can occur if non-standard settings are chosen; in orthorhombic C space groups, SL 23 can occur if the non-standard setting Ccma instead of Cmca is used.

According to the methods described above, it can be shown that some of the superlattices listed by Matsumoto and Wondratschek (1979) can not occur.

Example. Given a space group belonging to the crystal class mmm and conventional P lattice. A superlattice generated by the additional vector $\vec{v}_i = 1/4, 1/4, 0$ also contains the lattice vectors $1/4, 3/4, 0$; $3/4, 1/4, 0$; $3/4, 3/4, 0$. From equation (4) it can be seen that there are no solutions for the symmetry operations with rotation parts 1, 2_x, 2_y, m_x, m_y, and m_z. The remaining two symmetry operations $\bar{1}$ and 2_z (or 2_{1z}) are not sufficient to generate the four additional vectors of the superlattice.

Table 2. Superlattices of extraordinary crystallographic orbits
in 3-dimensional space.

Triclinic space groups

SL	1	2	P(1/2,1,1)	1/2,0,0
SL	2	2	P(1,1/2,1)	0,1/2,0
SL	3	2	P(1,1,1/2)	0,0,1/2
SL	4	2	A(1,1,1)	0,1/2,1/2
SL	5	2	B(1,1,1)	1/2,0,1/2
SL	6	2	C(1,1,1)	1/2,1/2,0
SL	7	2	I(1,1,1)	1/2,1/2,1/2

Monoclinic space groups

a) P lattice

SL	1	2	P(1/2,1,1)	1/2,0,0	
SL	2	2	P(1,1/2,1)	0,1/2,0	
SL	3	2	P(1,1,1/2)	0,0,1/2	
SL	4	2	A(1,1,1)	0,1/2,1/2	
SL	5	2	B(1,1,1)	1/2,0,1/2	
SL	6	2	C(1,1,1)	1/2,1/2,0	
SL	7	2	I(1,1,1)	1/2,1/2,1/2	
SL	9	4	P(1/2,1/2,1)	1/2,0,0	0,1/2,0
SL	10	4	P(1/2,1,1/2)	1/2,0,0	0,0,1/2
SL	11	4	A(1/2,1,1)	1/2,0,0	0,1/2,1/2
SL	14	4	I(1/2,1,1)	1/4,1/2,1/2	
SL	15	4	P(1,1/4,1)	0,1/4,0	
SL	16	4	P(1,1/2,1/2)	0,1/2,0	0,0,1/2
SL	17	4	A(1,1/2,1)	0,1/4,1/2	
SL	18	4	B(1,1/2,1)	0,1/2,0	1/2,0,1/2
SL	21	4	P(1,1,1/4)	0,0,1/4	
SL	22	4	A(1,1,1/2)	0,1/2,1/4	
SL	23	4	B(1,1,1/2)	1/2,0,1/4	
SL	24	4	C(1,1,1/2)	0,0,1/2	1/2,1/2,0
SL	25	4	I(1,1,1/2)	1/2,1/2,1/4	
SL	26	4	F(1,1,1)	0,1/2,1/2	1/2,0,1/2

b) C lattice

SL	1	2	P(1/2,1/2,1)	1/2,0,0	
SL	2	2	C(1,1,1/2)	0,0,1/2	
SL	3	2	F(1,1,1)	0,1/2,1/2	
SL	4	4	P(1/4,1/2,1)	1/4,0,0	
SL	5	4	P(1/2,1/4,1)	0,1/4,0	
SL	6	4	P(1/2,1/2,1/2)	1/2,0,0	0,0,1/2
SL	7	4	A(1/2,1/2,1)	0,1/4,1/2	
SL	8	4	B(1/2,1/2,1)	1/4,0,1/2	
SL	11	4	C(1,1,1/4)	0,0,1/4	
SL	12	4	F(1,1,1/2)	0,1/2,1/4	
SL	13	4	C(1,1,1/2*)	1/4,0,1/4	
SL	14	4	F(1,1,1*)	1/4,1/2,1/4	

Table 2 (continued)

Orthorhombic space groups

a) P lattice

SL	1	2	P(1/2,1,1)	1/2,0,0		
SL	2	2	P(1,1/2,1)	0,1/2,0		
SL	3	2	P(1,1,1/2)	0,0,1/2		
SL	4	2	A(1,1,1)	0,1/2,1/2		
SL	5	2	B(1,1,1)	1/2,0,1/2		
SL	6	2	C(1,1,1)	1/2,1/2,0		
SL	7	2	I(1,1,1)	1/2,1/2,1/2		
SL	8	4	P(1/4,1,1)	1/4,0,0		
SL	9	4	P(1/2,1/2,1)	1/2,0,0	0,1/2,0	
SL	10	4	P(1/2,1,1/2)	1/2,0,0	0,0,1/2	
SL	11	4	A(1/2,1,1)	1/2,0,0	0,1/2,1/2	
SL	12	4	B(1/2,1,1)	1/4,0,1/2		
SL	13	4	C(1/2,1,1)	1/4,1/2,0		
SL	14	4	I(1/2,1,1)	1/4,1/2,1/2		
SL	15	4	P(1,1/4,1)	0,1/4,0		
SL	16	4	P(1,1/2,1/2)	0,1/2,0	0,0,1/2	
SL	17	4	A(1,1/2,1)	0,1/4,1/2		
SL	18	4	B(1,1/2,1)	0,1/2,0	1/2,0,1/2	
SL	19	4	C(1,1/2,1)	1/2,1/4,0		
SL	20	4	I(1,1/2,1)	1/2,1/4,1/2		
SL	21	4	P(1,1,1/4)	0,0,1/4		
SL	22	4	A(1,1,1/2)	0,1/2,1/4		
SL	23	4	B(1,1,1/2)	1/2,0,1/4		
SL	24	4	C(1,1,1/2)	0,0,1/2	1/2,1/2,0	
SL	25	4	I(1,1,1/2)	1/2,1/2,1/4		
SL	26	4	F(1,1,1)	0,1/2,1/2	1/2,0,1/2	
SL	28	8	P(1/4,1/2,1)	1/4,0,0	0,1/2,0	
SL	29	8	P(1/4,1,1/2)	1/4,0,0	0,0,1/2	
SL	30	8	A(1/4,1,1)	1/4,0,0	0,1/2,1/2	
SL	34	8	P(1/2,1/4,1)	1/2,0,0	0,1/4,0	
SL	35	8	P(1/2,1/2,1/2)	1/2,0,0	0,1/2,0	0,0,1/2
SL	36	8	A(1/2,1/2,1)	1/2,0,0	0,1/4,1/2	
SL	37	8	B(1/2,1/2,1)	0,1/2,0	1/4,0,1/2	
SL	40	8	P(1/2,1,1/4)	1/2,0,0	0,0,1/4	
SL	41	8	A(1/2,1,1/2)	1/2,0,0	0,1/2,1/4	
SL	43	8	C(1/2,1,1/2)	0,0,1/2	1/4,1/2,0	
SL	45	8	F(1/2,1,1)	0,1/2,1/2	1/4,0,1/2	
SL	47	8	P(1,1/4,1/2)	0,1/4,0	0,0,1/2	
SL	49	8	B(1,1/4,1)	0,1/4,0	1/2,0,1/2	
SL	52	8	P(1,1/2,1/4)	0,1/2,0	0,0,1/4	
SL	54	8	B(1,1/2,1/2)	0,1/2,0	1/2,0,1/4	
SL	55	8	C(1,1/2,1/2)	0,0,1/2	1/2,1/4,0	
SL	57	8	F(1,1/2,1)	0,1/4,1/2	1/2,0,1/2	
SL	61	8	C(1,1,1/4)	0,0,1/4	1/2,1/2,0	
SL	63	8	F(1,1,1/2)	0,1/2,1/4	1/2,1/2,0	

Table 2 (continued)

b) C lattice

SL	1	2	P(1/2,1/2,1)	1/2,0,0	
SL	2	2	C(1,1,1/2)	0,0,1/2	
SL	3	2	F(1,1,1)	0,1/2,1/2	
SL	4	4	P(1/4,1/2,1)	1/4,0,0	
SL	5	4	P(1/2,1/4,1)	0,1/4,0	
SL	6	4	P(1/2,1/2,1/2)	1/2,0,0	0,0,1/2
SL	7	4	A(1/2,1/2,1)	0,1/4,1/2	
SL	8	4	B(1/2,1/2,1)	1/4,0,1/2	
SL	9	4	C(1/2,1/2,1)	1/2,0,0	1/4,1/4,0
SL	10	4	I(1/2,1/2,1)	1/2,0,0	1/4,1/4,1/2
SL	11	4	C(1,1,1/4)	0,0,1/4	
SL	12	4	F(1,1,1/2)	0,1/2,1/4	
SL	14	8	P(1/4,1/4,1)	1/4,0,0	0,1/4,0
SL	15	8	P(1/4,1/2,1/2)	1/4,0,0	0,0,1/2
SL	16	8	A(1/4,1/2,1)	1/4,0,0	0,1/4,1/2
SL	21	8	P(1/2,1/4,1/2)	0,1/4,0	0,0,1/2
SL	23	8	B(1/2,1/4,1)	0,1/4,0	1/4,0,1/2
SL	26	8	P(1/2,1/2,1/4)	1/2,0,0	0,0,1/4
SL	31	8	F(1/2,1/2,1)	0,1/4,1/2	1/4,1/4,0

c) A lattice

SL	1	2	P(1,1/2,1/2)	0,1/2,0	
SL	2	2	A(1/2,1,1)	1/2,0,0	
SL	3	2	F(1,1,1)	1/2,0,1/2	
SL	4	4	P(1/2,1/2,1/2)	1/2,0,0	0,1/2,0
SL	5	4	P(1,1/4,1/2)	0,1/4,0	
SL	6	4	P(1,1/2,1/4)	0,0,1/4	
SL	7	4	A(1,1/2,1/2)	0,1/2,0	0,1/4,1/4
SL	8	4	B(1,1/2,1/2)	1/2,0,1/4	
SL	9	4	C(1,1/2,1/2)	1/2,1/4,0	
SL	11	4	A(1/4,1,1)	1/4,0,0	
SL	12	4	F(1/2,1,1)	1/4,0,1/2	

d) I lattice

SL	1	2	A(1/2,1,1)	1/2,0,0	
SL	2	2	B(1,1/2,1)	0,1/2,0	
SL	3	2	C(1,1,1/2)	0,0,1/2	
SL	4	4	P(1/2,1/2,1/2)	1/2,0,0	0,1/2,0
SL	5	4	A(1/4,1,1)	1/4,0,0	
SL	6	4	F(1/2,1,1)	1/4,0,1/2	
SL	7	4	B(1,1/4,1)	0,1/4,0	
SL	8	4	F(1,1/2,1)	0,1/4,1/2	
SL	9	4	C(1,1,1/4)	0,0,1/4	
SL	10	4	F(1,1,1/2)	0,1/2,1/4	
SL	11	8	P(1/4,1/2,1/2)	1/4,0,0	0,1/2,0
SL	12	8	P(1/2,1/4,1/2)	1/2,0,0	0,1/4,0
SL	13	8	P(1/2,1/2,1/4)	1/2,0,0	0,0,1/4

Table 2 (continued)

e) F lattice

SL	1	2	P(1/2,1/2,1/2)	1/2,0,0		
SL	2	4	P(1/4,1/2,1/2)	1/4,0,0		
SL	3	4	P(1/2,1/4,1/2)	0,1/4,0		
SL	4	4	P(1/2,1/2,1/4)	0,0,1/4		
SL	5	4	A(1/2,1/2,1/2)	1/2,0,0	0,1/4,1/4	
SL	6	4	B(1/2,1/2,1/2)	1/2,0,0	1/4,0,1/4	
SL	7	4	C(1/2,1/2,1/2)	1/2,0,0	1/4,1/4,0	
SL	8	4	I(1/2,1/2,1/2)	1/4,1/4,1/4		
SL	9	8	P(1/8,1/2,1/2)	1/8,0,0		
SL	10	8	P(1/4,1/4,1/2)	1/4,0,0	0,1/4,0	
SL	11	8	P(1/4,1/2,1/4)	1/4,0,0	0,0,1/4	
SL	12	8	A(1/4,1/2,1/2)	1/4,0,0	0,1/4,1/4	
SL	17	8	P(1/2,1/4,1/4)	0,1/4,0	0,0,1/4	
SL	19	8	B(1/2,1/4,1/2)	0,1/4,0	1/4,0,1/4	
SL	25	8	C(1/2,1/2,1/4)	0,0,1/4	1/4,1/4,0	
SL	27	8	F(1/2,1/2,1/2)	1/2,0,0	0,1/4,1/4	1/4,0,1/4

Tetragonal space groups

a) P lattice

SL	1	2	P(1,1,1/2)	0,0,1/2		
SL	2	2	C(1,1,1)	1/2,1/2,0		
SL	3	2	I(1,1,1)	1/2,1/2,1/2		
SL	4	4	P(1,1,1/4)	0,0,1/4		
SL	5	4	C(1,1,1/2)	0,0,1/2	1/2,1/2,0	
S$^\mathsf{T}$	6	4	I(1,1,1/2)	1/2,1/2,1/4		
SL	7	4	P(1/2,1/2,1)	1/2,0,0	0,1/2,0	
SL	8	4	F(1,1,1)	0,1/2,1/2	1/2,0,1/2	
SL	9	8	P(1,1,1/8)	0,0,1/8		
SL	10	8	C(1,1,1/4)	0,0,1/4	1/2,1/2,0	
SL	11	8	I(1,1,1/4)	1/2,1/2,1/8		
SL	12	8	P(1/2,1/2,1/2)	1/2,0,0	0,1/2,0,0	0,1/2
SL	13	8	F(1,1,1/2)	1/2,1/2,0	1/2,0,1/4	
SL	14	8	C(1/2,1/2,1)	1/2,0,0	1/4,1/4,0	
SL	15	8	I(1/2,1/2,1)	1/2,0,0	1/4,1/4,1/2	
SL	19	16	P(1/2,1/2,1/4)	1/2,0,0	0,1/2,0	0,0,1/4
SL	21	16	C(1/2,1/2,1/2)	1/2,0,0	1/4,1/4,0	0,0,1/2

b) I lattice

SL	1	2	C(1,1,1/2)	0,0,1/2		
SL	2	4	C(1,1,1/4)	0,0,1/4		
SL	3	4	P(1/2,1/2,1/2)	1/2,0,0	0,1/2,0	
SL	4	4	F(1,1,1/2)	0,1/2,1/4		
SL	5	8	C(1,1,1/8)	0,0,1/8		
SL	6	8	P(1/2,1/2,1/4)	1/2,0,0	0,0,1/4	
SL	7	8	F(1,1,1/4)	0,1/2,1/8		
SL	8	8	C(1/2,1/2,1/2)	1/2,0,0	1/4,1/4,0	
SL	9	8	I(1/2,1/2,1/2)	1/2,0,0	0,1/2,0	1/4,1/4,1/4
SL	11	16	P(1/2,1/2,1/8)	1/2,0,0	0,0,1/8	
SL	13	16	C(1/2,1/2,1/4)	1/2,0,0	0,0,1/4	1/4,1/4,0

Table 2 (continued)

Trigonal space groups

a) P lattice

```
SL   1    2    P(1,1,1/2)       0,0,1/2
SL   2    3    P(1,1,1/3)       0,0,1/3
SL   3    3    P(1*,1*,1)       2/3,1/3,0
SL   4    3    RO(1,1,1)        2/3,1/3,1/3
SL   5    3    RR(1,1,1)        1/3,2/3,1/3
SL   6    4    P(1,1,1/4)       0,0,1/4
SL   8    6    P(1,1,1/6)       0,0,1/6
SL   9    6    P(1*,1*,1/2)     0,0,1/2   2/3,1/3,0
SL  10    6    RO(1,1,1/2)      2/3,1/3,1/6
SL  11    6    RR(1,1,1/2)      1/3,2/3,1/6
```

b) RO lattice

```
SL   1    2    RR(1,1,1/2)      0,0,1/2
SL   2    3    P(1*,1*,1/3)     0,0,1/3
SL   3    4    RO(1,1,1/4)      0,0,1/4
SL   5    6    P(1*,1*,1/6)     0,0,1/6
```

Hexagonal space groups

```
SL   1    2    P(1,1,1/2)       0,0,1/2
SL   2    3    P(1,1,1/3)       0,0,1/3
SL   3    3    P(1*,1*,1)       2/3,1/3,0
SL   4    4    P(1,1,1/4)       0,0,1/4
SL   6    6    P(1,1,1/6)       0,0,1/6
SL   7    6    P(1*,1*,1/2)     0,0,1/2   2/3,1/3,0
SL  10   12    P(1,1,1/12)      0,0,1/12
SL  11   12    P(1*,1*,1/4)     0,0,1/4   2/3,1/3,0
```

Cubic space groups

a) P lattice

```
SL   1    2    I(1,1,1)         1/2,1/2,1/2
SL   2    4    F(1,1,1)         0,1/2,1/2   1/2,0,1/2
SL   3    8    P(1/2,1/2,1/2)   1/2,0,0   0,1/2,0   0,0,1/2
```

b) I lattice

```
SL   1    4    P(1/2,1/2,1/2)   1/2,0,0   0,1/2,0
SL   2    8    I(1/2,1/2,1/2)   1/2,0,0   0,1/2,0   1/4,1/4,1/4
```

c) F lattice

```
SL   1    2    P(1/2,1/2,1/2)   1/2,0,0
SL   2    4    I(1/2,1/2,1/2)   1/4,1/4,1/4
SL   3    8    F(1/2,1/2,1/2)   1/2,0,0   0,1/4,1/4   1/4,0,1/4
SL   4   16    P(1/4,1/4,1/4)   1/4,0,0   0,1/4,0   0,0,1/4
```

5.2. The group theoretic approach

The group theoretic method proposed by Engel (1983) allows one to determine all non-characteristic orbits and to assign the eigensymmetry space group E to them. Let $O_G(X_0)$ be a non-characteristic orbit with $E > G$ and index $[E:G]=k$. In the generating space group G, X_0 has the site-symmetry group $S_G(X_0)$. In the eigensymmetry space group the point X_0 belongs to a special Wyckoff position with site-symmetry group $S_E(X_0)$. The ratio of the orders $h[S_E(X_0)]$ and $h[S_G(X_0)]$ of the site-symmetry groups is equal to the index k,

$$k = \frac{h[S_E(X_0)]}{h[S_G(X_0)]} \qquad (6)$$

Otherwise, the eigensymmetry space group E would generate more points than are contained in $O_G(X_0)$. The special Wyckoff positions in the eigensymmetry space group E which satisfy equation (6) are solutions for non-characteristic orbits of space group G.

If there exists another space group $\bar{E}$ between G and E, $G < \bar{E} < E$, then $O_{\bar{E}}(X_0)$ must also show the eigensymmetry space group $\bar{E}$ (The crystallographic orbit $O_E(X_0)$ cannot split into two crystallographic orbits when going from E to $\bar{E}$, because by definition $\bar{E}$ is also a supergroup of G). Hence, if there is no special Wyckoff position in $\bar{E}$ which satisfies equation (6) then the supergroups $E \geq \bar{E}$ cannot occur as eigensymmetry space groups. Therefore, all non-characteristic crystallographic orbits of G are already contained in the solutions of the minimal supergroups. If several groups $\bar{E}_i$, $i=1,...,m$ have a common supergroup $\bar{\bar{E}}$ then we find all solutions in $\bar{\bar{E}}$ through the intersection of the solutions of the $\bar{E}_i$. Finally, the eigensymmetry space group E is the highest supergroup. Usually there are several paths from the group G to the eigensymmetry group E; it is sufficient to establish one of them. An example is shown in Figure 2 for the space group No. 104, P4nc on p. 203.

In section 2.2 the affine normalizer $N_A(G)$ has been defined as the group of all affine mappings leaving G invariant. Analogously the Euclidean normalizer $N_{IE}(G)$ is the group of all isometries leaving G invariant. The Euclidean normalizers have been listed first by Hirshfeld (1968) under the name "Cheshire groups". Let E be a supergroup of G. In order to obtain all solutions for non-characteristic orbits with eigensymmetry E both Euclidean normalizers $N_{IE}(E)$ and $N_{IE}(G)$ are compared. Let be $G \le F = N_{IE}(G) \cap N_{IE}(E)$. If $F = N_{IE}(G)$ then we find in E all solutions M which are equivalent under $N_{IE}(G)$. However, if $F < N_{IE}(G)$ with index j then we determine the cosets of $N_{IE}(G)$ with respect to F,

$$N_{IE}(G) = F \cup FR_2 \cup \ldots \cup FR_j.$$

In this case we also have to consider, in addition to E, all supergroups $E_2 = R_2^{-1} ER_2, \ldots, E_j = R_j^{-1} ER_j,$ conjugate under $N_{IE}(G)$ in order to obtain all solutions. This means that for the solutions M in E, $R_2^{-1} M, \ldots, R_j^{-1} M$ are solutions in $E_2, \ldots, E_j$, but not in E. This case is shown in Figure 3 on p. 204. $P6_1 22$ is a supergroup of $P3_1 12$. However, for the Euclidean normalizers

$$N_{IE}(P3_1 12) = P6_2 22(1/3(\vec{a}-\vec{b}), 1/3(\vec{a}+2\vec{b}), 1/2\vec{c}) \text{ and}$$

$$N_{IE}(P6_1 22) = P6_2 22(1/2\vec{c})$$

$$F = P6_2 22(1/2\vec{c}) = N_{IE}(P6_1 22) \cap N_{IE}(P3_1 12)$$

holds (the basis vectors are given in parentheses). In this case $N_E(E) = F < N_E(G)$. Representatives of the coset decomposition are the translations

$$R_2 \sim 2/3, 1/3, 0 \qquad R_3 \sim 1/3, 2/3, 0.$$

Another example is space group No. 152, $P3_1 21$. For SL 4, $R\bar{3}m$ is a supergroup of $P3_1 21$. For the Euclidean normalizers $N_{IE}(R\bar{3}m) = R\bar{3}m(-\vec{b}, \vec{a}+\vec{b}, 1/2\vec{c})$ and $N_{IE}(P3_1 21) = P6_2 22(1/2\vec{c})$ we have

$$F = P3_2 21(1/2\vec{c}) = N_{IE}(R\bar{3}m) \cap N_{IE}(P3_1 21)$$

In this case $N_{IE}(E) > F < N_{IE}(G)$. A representative of the coset decomposition is the twofold rotation

$R_2 \sim 1-x, 1-y, z$.

This twofold rotation transforms the rhombohedral (obverse) RO lattice SL 4 into the (reverse) RR lattice SL 5. In essential the same manner group-subgroup and group-supergroup relations between space groups are discussed in a forthcoming paper by Koch (1984).

5.3. The calculation of non-characteristic crystallographic orbits

The extraorbits were calculated following the algebraic method described in section 5.1 using a program written by Engel in PL/I. The eigensymmetry space groups of the extraorbits and the non-characteristic orbits which have no additional translations were determined by hand by Engel using the Tables of minimal supergroups of the space groups (Neubüser and Wondratschek, 1969). These results were combined by hand. The Main Table has then been checked carefully by Engel, Matsumoto, Steinmann and Wondratschek and detected errors were corrected.

The control of the data has been performed in different ways.

(1) The non-characteristic orbits belonging to SL 0 have been checked in the same way as described in section 5.2. The table of lattice complexes by Fischer, Burzlaff, Hellner, and Donnay (1973) and later in IT 1983 have been found to be very useful here and in other steps of controlling the data.

(2) A given unit cell of a space group G is covered with a grid of such a small mesh width that each special position of a possible minimal supergroup is touched by at least one point of the grid. The orbit $O_G(X)$ from every point X of the grid is generated and subsequently tested for additional translations $\vec{v}_i$ (cf. section 5.1). The extraorbits detected are assigned to a special position of the eigensymmetry space group E using the group theoretic approach given in section 5.2.

(3) The non-characteristic orbits of superlattices with low density can be found by hand when considering the individual symmetry elements (distinguished points, axes, planes) as described in section 5.1. The combination of these results is easy in many cases and leads to the non-characteristic orbits with superlattices of higher density.

(4) Cross references were made in order to check the completeness of the Main Table. Let G be the generating space group and M a manifold of non-characteristic orbits with eigensymmetry space group E. In order to find all solutions M' equivalent to M the normalizer $N_E(G)$ has to be compared with E by considering the group $F=N_{IE}(G) \cap E$ (for orthorhombic space groups the affine normalizer $N_A(G)$ has been used). If $F=N_{IE}(G)$, then there is no solution equivalent to M. If $F<N_{IE}(G)$ of index j then the j-1 other solutions are obtained by decomposing $N_E(G)$ into cosets with respect to F and mapping M with the representatives of the cosets. In a similar way Fischer and Koch (1983) have treated the corresponding problem for characteristic orbits (E=G). There tables have been useful for checking the Main Table when they were available.

(5) The results for cubic space groups G have been checked by comparison with the corresponding orthorhombic or tetragonal maximal subgroups S. For each non-characteristic orbit of general position of G a corresponding solution in S must exist. The solutions of S thus lead to the solutions of G. The results for the general positions can be transferred to the special positions.

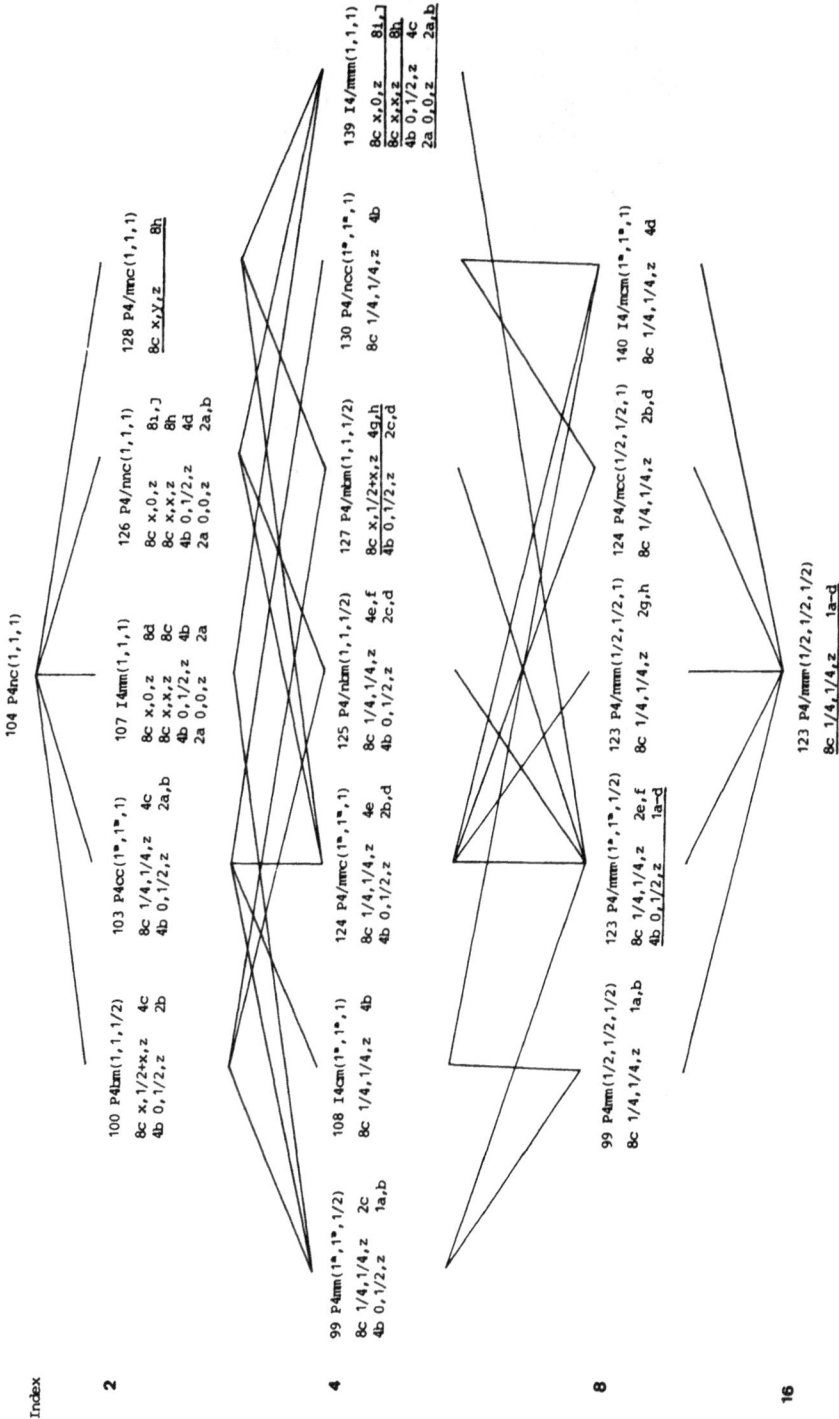

Figure 2. The non-characteristic crystallographic
orbits of the space group No. 104, P4nc.

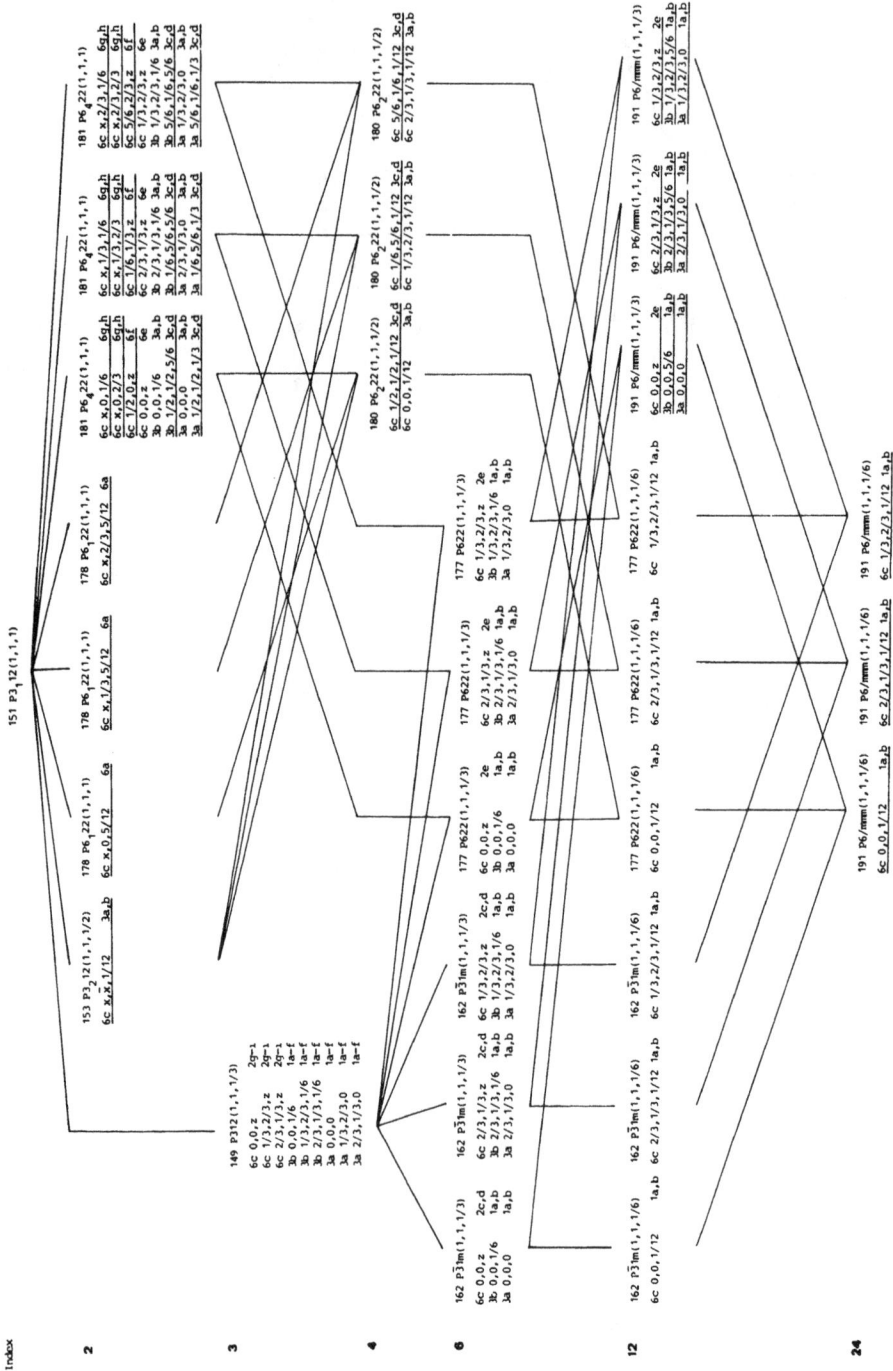

Figure 3. The non-characteristic crystallographic orbits of the space group No. 151, P3₁12.

6. Applications of non-characteristic crystallographic orbits

6.1. Extinction rules

The diffraction pattern of a crystal shows systematic extinctions depending on the space-group type of the crystal. 122 different extinction units, each containing up to five space-group types, may be distinguished by the general extinction rules. Extraorbits cause additional integral extinctions which are not required by the generating space group G but affect the whole diffraction pattern. Atoms in such positions contribute with their diffraction power to certain classes of reflections only. Such additional extinctions were investigated by Brandenberger (1928, 1929, 1931), Brandenberger and Niggli (1928) and Schneider (1931). However, their tables are incomplete.

If the projection of a crystallographic orbit onto a plane or a straight line shows additional translations then the contribution of this orbit to the diffraction pattern shows additional zonal or serial extinctions. Such projections were investigated by Moor (1983) and examples were given by him.

Non-characteristic crystallographic orbits may cause a special kind of diffraction enhancement (Sadanaga and Ohsumi, 1979) or may influence the intensity statistics.

6.2. Determination of the true space group

March and Herbstein (1983) have cited some examples where the stated space group of a crystal structure is only a subgroup of the true space group G . On the other hand, often for crystal chemical considerations a crystal-structure model is approximated with idealized atomic positions. In both cases the true space group can be obtained if for each atomic position X_i, i=1, 2, ..., N the eigensymmetry space group E_i of the corresponding crystallographic orbit is determined. The true space group is the intersection of the space groups E_i

$$G = E_1 \cap E_2 \cap \ldots \cap E_N.$$

Clearly G may be a proper subgroup of every of the eigensymmetry space groups.

Examples.
(1) In the NaCl structure , space group $Fm\bar{3}m$, the Na position 4a 0,0,0 has eigensymmetry $E_1 = Fm\bar{3}m$ and the Cl position 4b 1/2,1/2,1/2 has also eigensymmetry $E_2 = Fm\bar{3}m$. In this case $G = E_1 = E_2$.
(2) In the ZnS structure, space group $F\bar{4}3m$, the Zn position 4c 1/4,1/4,1/4 has eigensymmetry $E_1 = Fm\bar{3}m$ and the S position 4a 0,0,0 has also eigensymmetry $E_2 = Fm\bar{3}m$. In this case not all symmetry operations of E_1 and E_2 are retained as symmetry operations of the crystal structure and thus $G < E_1$, $G < E_2$ holds, though E_1 and E_2 belong to the same space-group type.
(3) In the CaF_2 structure, space group $Fm\bar{3}m$, the Ca position 4a 0,0,0 has eigensymmetry $E_1 = Fm\bar{3}m$ and the F position 8c 1/4,1/4,1/4 has eigensymmetry $E_2 = Pm\bar{3}m$, $a' = 1/2a$. In this case $G = E_1 < E_2$.
(4) In the Cu_2O structure, space group $Pn\bar{3}m$, the Cu position 4b 0,0,0 has eigensymmetry $E_1 = Fm\bar{3}m$ and the O position 2a 1/4,1/4,1/4 has eigensymmetry $E_2 = Im\bar{3}m$. In this case $G = Pn\bar{3}m$, $G < E_1$, $G < E_2$, and $E_1 \neq E_2$; E_1 and E_2 belong to different space-group types.

6.3 Phase transitions

If the symmetries of two crystal structures are in a group-subgroup relation of index k, then the high-symmetry structure may be obtained by an appropriate displacement of the atoms of the low-symmetry structure (if crystallographic orbits are substituted for the atoms, by specialized coordinates of the orbits) and/or by additional restrictions for the lattice parameters.

The symmetry of a crystal structure may be increased by the following mechanisms:

(1) A characteristic crystallographic orbit may change to become a non-characteristic orbit or a non-characteristic orbit may change to become another non-characteristic orbit. In both cases the order of the site symmetry of the crystallographic orbit in the new space group is enhanced by the index k.

(2) k crystallographic orbits may combine to form a
    new crystallographic orbit, the site symmetry
    being unchanged.
(3) Any combination of (1) and (2), i.e. j crystal-
    lograhic orbits, j<k, may combine to form a new
    crystallographic orbit, the site symmetry of the
    points from at least one of these crystallo-
    graphic orbits is increased.

Using the Main Table one can obtain information
about mechanism (1).

Example: Transition of sodium niobate at 600°C from
         tetragonal to cubic perovskite structure
         (Glazer and Megaw, 1972)

From the data on subgroups in IT 1983 one obtains
the group-subgroup relations (lattice constants in
parentheses): P4/mbm($\sqrt{2}$a,$\sqrt{2}$b,c) -> P4/mmm(a,b,c)
-> Pm$\bar{3}$m(a,b,c). The structure of lower symmetry
(space group P4/mbm) is represented by the crystallographic orbits:

Nb	2 a	4/m	0,0,0
Na	2 c	mmm	0,1/2,1/2
O(1)	2 b	4/m	0,0,1/2
O(2)	4 g	mm2	x,1/2+x,0; (x ~ 1/4)

One can obtain the non-characteristic crystallographic orbits directly (without cell transformation) from the Main Table.

2 a	4/m	SL 2	0,0,0	P4/mmm	(1a-d)
2 c	mmm	SL 2	0,1/2,1/2	P4/mmm	(1a-d)
2 b	4/m	SL 2	0,0,1/2	P4/mmm	(1a-d)
4 g	mm2	SL 7	1/4,3/4,0	P4/mmm	(1a-d)

Obviously the crystallographic orbits 2a, 2b and 2c
of space group P4/mbm show eigensymmetry P4/mmm and
SL 2. The crystallographic orbit 4g will have a
higher eigensymmetry only for x=1/4. SL 7 is a
superlattice of SL 2, the intersection of the eigensymmetry space groups of the crystallographic orbits
changes from P4/mbm($\sqrt{2}$a,$\sqrt{2}$b,c) to P4/mmm(a,b,c).

The second step P4/mmm -> Pm$\bar{3}$m cannot be found in
the Main Table because both space groups belong to
different crystal families. Therefore, "accidental"
lattice symmetry is required which is not accounted

for in the Main Table. The relation can be determined using the method described in section 5.2.

6.4. Generation of a crystallographic orbit

The Main Table may also be helpful in obtaining an answer to the question: Find all space groups and all coordinates from which a given crystallographic orbit is generated (without composition from more than one crystallographic orbit).

Examples.
(1) In the idealized perovskite structure ABO_3, space group $Pm\overline{3}m$, the positions of the oxygen atoms are described by the crystallographic orbit 3c (or 3d). In Table 1 all cubic space groups are listed which can generate this crystallographic orbit type $Pm\overline{3}m$ (3c,d). In those space groups where the Wyckoff letter is preceded by an "ex", the crystallographic orbit may be distorted according to the free parameters. Doing this its eigensymmetry is reduced but the equivalent points remain equivalent.
(2) Sandor (1968) proposed to list the coordinates of all points in the unit cell for which the crystallographic orbits have "higher translational symmetry than the general position", i.e. to list all extraordinary orbits. He tabulates the occurence of the orthorhombic F lattice in non-face-centred orthorhombic space groups. A comparison of his list with the Main Table shows, however, that his Table 2 contains less than half the relevant data, whereas 9 from his 62 cases should be cancelled due to equivalence.
(3) Chuprunov, Tarkhova, Talis, Suvorova, and Belov (1981) list the extraordinary orbits of the triclinic and monoclinic space groups. Several errors occur in their data for the space-group types No. 12, B2/m to No. 15, B2/b (unique axis c).

7. Historical remarks

The investigation of space groups was greatly stimu-
lated by the crystalline state of matter. Johann
Keppler(1571-1630) and Robert Hooke (1635-1703) were
among the first to try to explain the regular shapes
of crystals by a lattice-like arrangement of bullets
or globular particles. These ideas were further
developed by Christian Huygens (1629-1695) and nota-
bly by René Just Haüy (1743-1822) who proposed
parallelepipedal building bricks called "molécules
soustractives" which themselves were constructed by
"molécules intégrantes".

Today these concepts of sphere and ellipsoid packings and of space tilings generate renewed interest
and are still useful geometrical tools for the
understanding of crystal structures. We mention here
the important work of Niggli (1927, 1928b) on sphere
packings. These concepts were further investigated
by Heesch and Laves (1931), Sinogowitz (1939, 1943)
and by Fischer (1971,1973,1974b). The ellipsoid
packings in the plane were investigated by Nowacki
(1948) and Grünbaum and Shephard (1979). Nowacki and
Matsumoto (1966) investigated some ellipsoid packings in three-dimensional space. Space tilings were
derived by Laves (1931), Nowacki (1935), Löckenhoff
and Hellner (1971), Koch (1972) and Engel (1981a,
1981b).

An atomistic theory was introduced by the physicist
Seeber (1824) who postulated small spherical atoms
which remain in a stable equilibrium by the balance
of attractive and repulsive forces. These atoms were
assumed to be situated at the nodes of a space
lattice.

Independently Delafosse (1843), considering only the
centres of gravity of the molecules, argued that
these centres have to form a point lattice. In other
words the centres must lie at the nodes of three
mutually intersecting sets of equidistant parallel
planes.

These ideas of Seeber and Delafosse opened a new
area in crystallography: the investigation of regular point sets. At that time it was assumed that
the molecules in a crystal structure would have to
be arranged parallel in space and thus the mass

centres would form a point lattice. Point lattices
may be classified according to their symmetry. In
this sense, Frankenheim (1842) found 15 types of
point lattices in 3-dimensional space. But Bravais
(1850) proved that only 14 different lattice types,
now known as the 14 Bravais lattices, exist.

An extension of this arbitrary assumption of point
lattices was introduced by Wiener (1863) and by
Sohncke (1867). Following Sohncke, an infinite
discrete point system is one which shows the same
distribution of points about each of its points.
Later Sohncke (1874) gave a more rigorous definition:

   A discrete point system is regular if from any
   two points of the system straight lines are
   drawn to all the other points of the system
   and these two line systems are directly or
   symmetrically congruent.

Remarkably this definition of a regular point system
relies on purely geometrical evidence and does not
refer to space groups.

Sohncke (1874) investigated the regular point
systems in the plane and found 13 different
construction principles. These correspond exactly to
the 13 posssible types of eigensymmetry plane
groups. By specializing the construction conditions
Sohncke also found, for each construction principle,
all the possible eigensymmetry plane groups. These
results are summarized in Table 3.

Table 3. Sohncke's 13 construction types and their
possible eigensymmetry plane groups.

plane group type	Sohncke's construction type	possible eigensymmetry plane groups
p2	XI	p2, p2mm, p2mg, c2mm, p4mm, p6mm
p2mg	XII	p2mg, p2mm, c2mm, p4mm, p6mm
p2gg	XIII	p2gg, p2mm, c2mm, p4mm, p6mm
p2mm	X	p2mm, p4mm
c2mm	VI	c2mm, p2mm, p4mm, p6mm
p4	IV	p4, p4mm
p4mm	VII	p4mm
p4gm	V	p4gm, p4mm
p3	III	p3, p3m1, p31m, p6mm
p3m1	IX	p3m1, p6mm
p31m	II	p31m, p6mm
p6	I	p6, p6mm
p6mm	VIII	p6mm

The underscored plane groups are documented by
figures in Sohncke's paper.

At that time the determination of the space-group
types by Schönflies (1891) and simultaneously by
Fedorov (1892) interrupted the geometrical investigation of regular point sets. These investigations
were taken up again by Niggli (1919, 1928a) when he
introduced the concept of lattice complex. In our
opinion, Niggli (1919, p. 135, 414ff, 1928a, p. 201,
219) used "lattice complex" synonymously to what is
called "crystallographic orbit" here.   Another
interpretation was given by Burzlaff, Fischer, Hellner and A.Niggli (1974), they follow the concept of
Hermann (1935). P. Niggli showed that the same lattice
complex may occur in various space-group types. Thus
the eigensymmetry space group of the "lattice
complex" may be a supergroup of the generating space
group. In 1919 Niggli wrote: "in future one of the
most important problems in crystallography will be
to accomplish the Table of lattice complexes in this
respect".

However, it was only recently that further progress
in the investigation of regular point systems was
achieved.   (Wondratschek,   1976;   Lawrenson   and

Wondratschek, 1976; Matsumoto and Wondratsckek 1979; Engel, 1983).

The concept of lattice complex was developed in a different way by Hermann (1935,1960); Menzer (1960); Fischer, Burzlaff, Hellner and Donnay (1973); Burzlaff, Fischer, Hellner and Niggli (1974); Fischer and Koch (1974a); Zimmermann and Burzlaff (1974). Extensive tables of lattice complexes are available in IT 1983.

Acknowledgements

This work was supported substantially by the
Deutsche Forschungsgemeinschaft through research
grants given to H. Wondratschek. The contribution of
H. von Benda to the earlier stages of the project is
gratefully acknowledged as well as the advice and
helpful remarks of J. J. Burckhardt and M. Senechal.
The collaboration with T. Matsumoto was possible
through the generous financial support by the Alex-
ander von Humboldt-Stiftung in Germany, the Japan
Society for Promotion of Science, and the Ministry
of Education, Science, and Culture in Japan. The
computations and the printing were done at the
computer centre of the University Berne (BEDAG). We
thank the Verlag Oldenbourg for offering to publish
the volume as a supplement to Zeitschrift für Kristallographie and for the efforts to keep the price
of this volume as low as possible.

References

Billiet, Y., Burzlaff, H. and Zimmermann, H.: Comment on the paper of H. Burzlaff and H. Zimmermann "On the choice of origin in the description of space groups". Z. Kristallogr. 160 (1982) 155-157

Brandenberger, E.: Systematische Darstellung der kristallstrukturell wichtigen Auswahlregeln trikliner, monokliner und rhombischer Raumsysteme. Z. Kristallogr. 68 (1928) 330-362

Brandenberger, E.: Systematische Darstellung der kristallstrukturell wichtigen Auswahlregeln tetragonaler Raumsysteme. Z. Kristallogr. 71 (1929) 452-500

Brandenberger, E.: Auswahlregeln, erzeugende Operationen und zugehörige Punktmannigfaltigkeiten der Kristallstrukturen. Z. Kristallogr. 76 (1931) 1-86

Brandenberger, E. und Niggli, P.: Die systematische Darstellung der kristallstrukturell wichtigen Auswahlregeln. Z. Kristallogr. 68 (1928) 301-329

Bravais, A.: Mémoire sur les systèm formés par des points distribués régulièrement sur un plan ou dans l'éspace. Journal de l'école polytechnique, T. 19 (1850) 1-128.

Burzlaff, H., Fischer, W., Hellner, E. und Niggli, A.: Zur Entwicklung des Begriffs "Gitterkomplex". Z. Kristallogr. 139 (1974) 246-251

Burzlaff, H. and Zimmermann, H.: On the choice of origins in the description of space groups. Z. Kristallogr. 153 (1980) 151-179

Chuprunov, E.V., Tarkhova, T.N., Talis, A.L., Suvorova, G.F., and Belov, N.V.: Pseudotranslational positions in Fedorov groups of the lower crystal systems. Kristallografiya 26 (1981) 5-7

Delafosse, G.: Recherches sur la cristallisation considérée sous les rapports physiques et mathématiques. Mémoire des savants étrangèrs, T. 8, Paris (1843).

Engel, P.: Ueber Wirkungsbereichsteilungen von kubischer Symmetrie. Z. Kristallogr. 154 (1981a) 199-215

Engel, P.: Ueber Wirkunsbereichsteilungen von kubischer Symmetrie. II. Die Typen von Wirkungsbereichspolyedern in den symmorphen kubischen Raumgruppen. Z. Kristallogr. 157 (1981b) 259-275

Engel, P.: Zur Theorie der kristallographischen Orbits. Z. Kristallogr. 163 (1983) 243-249

Fedorov, E. S.: Zusammenstellung der kristallographischen Resultate des Herrn Schönflies und der meinigen. Z. Kristallogr. 20 (1892) 25-75

Fischer, W.: Existenzbedingungen homogener Kugelpackungen in Raumgruppen tetragonaler Symmetrie. Z. Kristallogr. 133 (1971) 18-42

Fischer, W.: Existenzbedingungen homogener Kugelpackungen zu kubischen Gitterkomplexen mit weniger als drei Freiheitsgraden. Z. Kristallogr. 138 (1973) 129-146

Fischer, W., Burzlaff, H., Hellner, E., Donnay, J. D. H.: Space groups and lattice complexes. Washington, National Bureau of Standards Monograph 134 (1973)

Fischer, W. und Koch, E.: Eine Definition des Begriffs "Gitterkomplex". Z. Kristallogr. 139 (1974a) 268-278

Fischer, W.: Existenzbedingungen homogener Kugelpackungen zu kubischen Gitterkomplexen mit drei Freiheitsgraden. Z. Kristallogr. 140 (1974b) 50-74

Fischer, W. and Koch, E.: On the equivalence of point configurations due to Euclidean normalizers (Cheshire groups) of space groups. Acta Crystallogr. A39 (1983) 907-915

Frankenheim, M. L.: System der Crystalle. Nova Acta Acad. Caesareae Leopoldino-Carolinae Naturae Curiosorum. 19 (1842) 471-660

Glazer, A. M. and Megaw, H. D.: The structure of sodium niobate (T_2) at 600°C, and the cubic-tetragonal transition in relation to soft-phonon modes. Phil. Mag. 25 (1972) 1119-1135

Grünbaum, B. and Shephard, G. C.: Refinements of the crystallographic classification. Symposium on Mathematical Crystallography, Riederalp (1979)

Gubler, M.G.: Ueber die Symmetrien der Symmetriegruppen: Automorphismengruppen, Normalisatorgruppen und charakteristische Untergruppen von Symmetriegruppen, insbesondere der kristallographischen Punkt- und Raumgruppen. Diss. Univ. Zürich (1982)

Hahn, Th.: International Tables , Vol. A: Space-Group Symmetry. D. Reidel Publishing Company, Dordrecht (1983)

Heesch, H. and Laves, F.: Ueber dünne Kugelpackungen. Z. Kristallogr. 85 (1931) 443-453.

Hermann, C.: Zur systematischen Strukturtheorie. IV. Untergruppen. Z. Kristallogr. 69 (1929) 533-555

Hermann, C.: Internationale Tabellen zur Bestimmung von Kristallstrukturen. Band I, Gruppentheoretische Tafeln. Gebrüder Bornträger, Berlin (1935)

Hermann, C.: Zur Nomenklatur der Gitterkomplexe. Z. Kristallogr. 113 (1960) 142-154

Hirshfeld, F.L.: Symmetry in the generation of the trial structures. Acta Crystallogr. A24 (1968) 301-311

IT 1983 (see Hahn, 1983)

Koch, E.: Wirkungsbereichspolyeder und Wirkungsbereichsteilungen zu kubischen Gitterkomplexen mit weniger als drei Freiheitsgraden. Diss. Univ. Marburg (1972)

Koch, E.: Die Grenzformen der kubischen Gitterkomplexe. Z. Kristallogr. 140 (1974) 75-86

Koch, E.: The implications of normalizers on group-subgroup relations between space groups. Acta Crystallogr. (1984) to appear.

Laves, F.: Ebenenteilung in Wirkungsbereiche. Z. Kristallogr. 76 (1931) 277-284

Lawrenson, J. E. and Wondratschek, H.: The extraordinary orbits of the 17 plane groups. Z. Kristallogr. 143 (1976) 471-484

Löckenhoff, H. D. und Hellner, E.: Die Wirkungsbereiche der invarianten kubischen Gitterkomplexe. N. Jb. Miner. Mh. (1971) 155-179

March, R. E. and Herbstein, F. H.: Some additional changes in space groups of published crystal structures. Acta Crystallogr. B39 (1983) 280-287

Matsumoto, T. and Wondratschek, H.: Possible superlattices of extraordinary orbits in 3-dimensional space. Z. Kristallogr. 150 (1979) 181-198

Moor, R.: Auslöschungs-Grenzpunktlagen und ihre Bedeutung für die Kristallstrukturlehre. Diss. ETH Zürich (1983)

Menzer, G.: Symbole der Gitterkomplexe. Z. Kristallogr. 113 (1960) 178-194

Neubüser, J. and Wondratschek, H.: Maximal subgroups of the space groups (1969), unpublished

Niggli, P.: Geometrische Kristallographie des Diskontinuums. Gebrüder Bornträger, Leipzig (1919)

Niggli, P.: XXIV. Die topologische Strukturanalyse. I. Z. Kristallogr. 65 (1927) 391-415

Niggli, P.: Handbuch der Experimentalphysik: Kristallographische und Strukturtheoretische Grundbegriffe. Akad. Verlagsgesellschaft, Leipzig (1928a)

Niggli, P.: XXV. Die topologische Strukturanalyse. II. Z. Kristallogr. 68 (1928b) 404-466

Nowacki, W.: Homogene Raumteilung und Kristallstruktur. Diss. ETH Zürich (1935)

Nowacki, W.: Ueber Ellipsenpackungen in der Kristallebene. Schweiz. Min. Petr. Mitt. 28 (1948) 502-508

Nowacki, W. and Matsumoto, T.: On densest packings of ellipsoids. Z. Kristallogr. 123 (1966) 401-421

Nowacki, W.: Die scheinbaren Translationsgitter der Raumgruppen. N. Jb. Miner. Mh. (1975) 526-528

Sadanaga, R. and Ohsumi, K.: Basic theorems of vector symmetry in crystallography. Acta Crystallogr. A35 (1979) 115-122

Sandor, E.: Special positions in space groups. Z. Kristallogr. 126 (1968) 277-281

Schneider, E.: Systematische Darstellung der kristallstrukturell wichtigen Auswahlregeln hexagonaler und rhomboedrischer Raumsysteme. Z. Kristallogr. 77 (1931) 275-316

Schönflies, A.: Krystallsysteme und Krystallstructur. Verlag B. G. Teubner, Leipzig, (1891)

Seeber, L. A.: Versuch einer Erklärung des innern Baus der festen Körper. Gilberts Annalen der Physik, 76 (1824) 229-248

Sinogowitz, U.: Die Kreislagen und Packungen kongruenter Kreise in der Ebene. Z. Kristallogr. 100 (1939) 461-508

Sinogowitz, U.: Herleitung aller homogenen nicht kubischen Kugelpackungen. Z. Kristallogr. 105 (1943) 23-52

Sohncke, L.: Die Gruppirung der Molecüle in den Krystallen. Poggendorffs Ann. d. Phys. 132 (1867) 75-106

Sohncke, L.: Die regelmässigen ebenen Punktsysteme von unbegrenzter Ausdehnung. Journal für die reine und angewandte Math. 77 (1874) 47-101

Wiener, Chr.: Grundzüge der Weltordnung. Leipzig und Heidelberg (1863)

Wondratschek, H.: Extraordinary orbits of the space groups. Theoretical considerations. Z. Kristallogr. 143 (1976) 460-470

Zimmermann, H. und Burzlaff, H.: Zur Definition des Punktlagen- und Gitterkomplex- Begriffs. Z. Kristallogr. 139 (1974) 252-267

Correction and notes (June 1984)

p.6, first section, insert after line 3:
The symbols for Wyckoff sets or types of Wyckoff
sets as applied here differ from those used by,
e.g., Hermann (1935) and IT 1983. The present
symbols give more information without being more
complicated.

----

p. 192, line 1:
replace "translation" by "translation vector".

----

p. 202, add:

(6) The earlier unpublished results of H. v. Benda
    (extraorbits for about half the space-group
    types) and T. Matsumoto and H. Wondratschek
    (extraorbits of triclinic to orthorhombic space
    groups, exept those of crystal class mmm) have
    been compared with these tables. Differences,
    when occuring, have beeen eliminated.

(7) E. Koch (1974) lists the limiting complexes of
    the cubic lattice complexes. Her group-theore-
    tical method, cf. also Fischer, W. and Koch, E.:
    Limiting forms and comprehensive complexes for
    crystallographic point groups, rod groups and
    layer groups. Z. Kristallogr. 147 (1978)
    255-273, is similar to that applied in section
    5.2. The results have been used to check the
    corresponding entries of these tables.

(8) After the manuscript had been finished the paper
    of Koch, E.: A geometrical classification of
    cubic point configurations. Z. Kristallogr. 166
    (1984) 23-52, has appeared. Among other data the
    eigensymmetries of certain non-characteristic
    orbits of cubic space groups are given which
    have been compared with the corresponding eigen-
    symmetries of these tables.

(9) W. Fischer and E. Koch, private communication,
    have drawn the authors' attention to their
    paper: Kubische Strukturtypen mit festen Koordi-
    naten. Z. Kristallogr. 140 (1974) · 324-330. In
    this paper those non-characteristic cubic orbits
    from Wyckoff positions with free parameters are
    listed whose eigensymmetries belong to Wyckoff
    sets with fixed coordinates. Two misprints of
    the present tables have been eliminated when
    comparing the data. Moreover, Fischer and Koch
    l.c. give examples of crystal structures which
    are fully or partly built up by such crystallo-
    graphic orbits.

www.ingramcontent.com/pod-product-compliance
Lightning Source LLC
Chambersburg PA
CBHW081539190326
41458CB00015B/5596